W0254376

Science for Survival

Scientific Research and the Public Interest

Science for Survival

Scientific Research and the Public Interest

PETER COTGREAVE

THE BRITISH LIBRARY

First published 2003 by
The British Library
96 Euston Road
London NW1 2DB

British Library Cataloguing in Publication Data
A catalogue record for this book is available from the British Library

ISBN 0-7123-0891-1

Designed by Bob Elliott
Typeset by Hope Services (Abingdon) Ltd
Printed in England by Biddles, Guildford

Contents

Acknowledgements

ALTHOUGH I have written this book in a personal capacity, it would be facile to pretend that it had not been heavily influenced by my work with the campaigning organisation known as the Save British Science Society (SBS). In particular, my views have been developed in discussions with the Chairman of SBS, Professor Richard Joyner, and other members of the Executive Committee and Advisory Council, and with the many government ministers, businessmen, civil servants, journalists, campaigners, academics and others with whom I have interacted as part of my job.

I have, of course, had access to the archives and reference collection of SBS, and have drawn on material that was originally prepared for other purposes. In doing so, I have relied on research carried out by my colleague Alice Sharp Pierson, and I am grateful to her for the time and care she has put into generating this invaluable resource, and for expressing her forthright opinions about the issues that are contained in this book.

As an adult, it would be easy to forget the route that led to my experiences and opinions, but I know that I have only been able to follow an interesting and rewarding career because my parents, both of whom left school at 16, have always taken the trouble to be interested in what I have done, and to support me even though I have yet to get a 'proper job'.

Scientists are, in general, rightly sceptical of books that cite newspapers as primary sources, but I make no apology for having cited many newspapers from the British Library's holdings, because this is not a scientific tome, but a book about how society interacts with science and with the ways in which science is done. Because it touches on scientific issues and on public policy, I have drawn heavily on the British Library's collections in these areas, and I have also cited books and documents from other parts of the Library's impressive holdings. The staff in the various reading rooms have always been friendly and helpful, and have almost always been able to find what I wanted quickly and simply.

At the British Library, Anthony Warshaw has followed an exemplary course as a publisher. He gently and politely pestered me until I agreed to write the book, and then got me to sign a contract almost immediately, after which he let me get on with it. Last, I should like to thank Cathy

Cooper, who has convincingly pretended to be interested in the book, and whose comments have been extremely useful in improving the text. I claim exclusive credit for the book's many deficiencies.

Introduction

SCIENCE is everywhere. It touches all parts of our everyday lives, and the ever-changing technology that it delivers is the fundamental basis of our economy. We cannot escape from science and engineering, as they appear more and more on the television, as they take up an increasing number of pages in the newspapers and on the internet, and as the number of popular science books grows too rapidly for us to keep pace.

Science and technology have delivered unquestionable benefits that have made our lives happier, healthier and easier. But they also bring risks, and many of us are sceptical about some of the supposed benefits. One of the reasons that science hits the headlines so often is that an increasing number of citizens are questioning the value of some scientific progress. Not everybody is keen to see genetically-modified foods on our tables, and many of us question the value of adding fluoride to drinking water or the desirability of relying too heavily on nuclear power.

Others try to give the impression that scientific research is a great panacea that can continually deliver more and more for the peoples of the world. These scientific enthusiasts berate politicians and officials for not being scientifically literate, and bemoan the fact that we sometimes have a negative image of scientists. Whether you want to criticize science or to praise it, you can no longer ignore it. Whichever camp you belong to, there is no denying that society needs to discuss science much more than it used to do.

Many members of the scientific community have, in recent years, taken the view that this new and expanded dialogue should concentrate directly on the results of scientific inquiry. They believe that the solution to any arguments (between those in favour of particular scientific advances on the one hand and those who are sceptical about them on the other) is for the scientists to proselytize about their subject, which they do in two ways.

One group focuses on informing non-specialists about the wonders of some aspect of scientific study, such as the formation of stars, or the intricate beauty of the genetic code. This used to be the basis for the 'public understanding of science', but the phrase is falling out of fashion. This is because most scientists realize that it is unhelpful to give the impression that scepticism on the part of laypeople is due solely to the fact that they do not understand science.

Another group emphasizes how important scientific advancement has been in making our lives happier, healthier and easier. They point out the value of all the gadgets that modern life requires: digital television, the internet, mobile phones, ever more powerful drugs, and the bewildering array of controls on a car dashboard that will do anything from telling you where the traffic is heaviest to calming your nerves by playing a Mozart opera while you wait in a traffic jam (although a sceptic might point out that there would be no need for the calming influence of the music if the gizmo for providing traffic information had worked properly).

In general, the combined efforts of these two sets of scientific evangelists have been very successful. Publishers produce new popular science books faster than they produce almost any other titles, and television producers make ever more amazing science programmes. They are producing these books and programmes because people want them, and because we are prepared to pay for them.

These efforts to excite the public about science are an essential and integral part of a modern society and, without them, any debate about the place of science in society would be incomplete, because the majority of non-scientists would feel hugely disfranchised. However, it is not enough to excite some people, or even to engage them in scientific activity, as many scientists aim to do, with books, broadcasts, visitor attractions, educational schemes and internet sites.

Many people still do not think that they need to worry about how science is done, or how it is organized and funded, believing these issues to be matters largely for the scientific community or, where public money is involved, for their government or public officials. This applies even to many of those people who read popular science books, avidly view science documentaries, and frequently visit the local zoo, planetarium, or science museum whenever they arrive in a new town. The mechanics of science, the research itself, goes on behind the scenes, and is not of any interest to many members of the public.

But the public interest is different from the things in which the public expresses an interest. We live in a world where people in all walks of life, and especially those who receive public funding, are increasingly accountable, and in which an ever-expanding media helps the public to hold governments to account on a whole range of issues.[1]

[1] During the week of 1 February 1998, 24% of the UK-based news stories in the main broadsheet newspapers in the UK were about public spending. The average amount of money involved in each

It is highly undesirable that even those who care passionately about politics or about public policies rarely care about the politics of science. It is damaging from the point of view both of those who wish public support for science to increase, and of those who are sceptical about what they perceive as blanket claims that science will deliver a continuous stream of benefits without introducing significant problems.

Those who wish to lobby for science must appreciate that in the modern age of spin doctors and focus groups, public funding for science will only keep pace with scientific opportunities if our political leaders believe that the public broadly supports such a policy.

Moreover, governments will only deliver the funding structures needed to maximize scientific achievement if they perceive a large enough groundswell of public support. This is particularly true of support for the kind of fundamental science that is often referred to as 'blue skies' research, and which has no obvious and immediate purpose, but which contributes both to our cultural life and to the body of knowledge from which useful discoveries and inventions are drawn.

Evidence that politicians and their advisers do not always take scientists seriously could not be more graphic than the words of an anonymous aide to President Bill Clinton, who told the writer Daniel Greenberg that dealing with scientists was 'like talking to mental patients. You have to look like you take them seriously'.[2]

Greenberg went on to compile a few statistics that shed light on the low level of importance that the American political system has given to its relationship with science. In a 793-page study of contemporary politics, Hedrick Smith, former Washington bureau chief of the *New York Times*, did not mention science at all, apart from a few passages about the Strategic Defense Initiative, an issue that was simmering at the time he was writing in 1988. In his memoirs, President Lyndon Johnson did not even mention his personal science adviser, Donald Hornig, who served him inside the White House for five years. Richard Nixon's monumental 1090-page biography makes no mention of his declaration of 'war on cancer' or of his abolition of the White House Office of Science and Technology Policy.[3]

---

story was approximately £200 million, but the lowest was just £24,000. The newspapers included in the sample were: *The Daily Telegraph*, *The Guardian*, *The Independent* and *The Times*.

[2] Reported in Shapin, S. (2001) Guests in the President's House. *London Review of Books*, 18 October 2001, p.7.

[3] Greenberg, D.S. (2001) *Science, Money and Politics: Political Triumph and Ethical Erosion*, p.345. University of Chicago Press, Chicago.

The same type of analysis produces similar results in other countries. The index to a 400-page book about the lives and policies of British prime ministers contains only one reference to science, and that refers to the 1920s when Arthur Balfour created the Medical Research Council.[4] It is no wonder that advocates and lobbyists for science feel that more should be done to engage political opinion with those issues of public interest that have a scientific dimension.

But the failure of the scientific and governmental communities to explain the organization and funding structures that they have adopted is equally unhelpful to those who wish to campaign against particular aspects of scientific research. There is little point, for example, in attempting to convince people that the government must control research on the genetic modification of crops if your audience has not appreciated the fundamental difference between government support that comes through the relevant policy-oriented agricultural ministry, and government support that comes through a Research Council, charged merely with seeking out the best and most groundbreaking science.

Anyone opposing scientific developments, or arguing that new technologies are against the public interest, can increase the impact of their case by supporting it with an appreciation of how the science is organized and funded.

In the past few years, the 'public understanding of science' has largely been replaced in the minds of the science community by a debate about 'science and society' (or more accurately 'science in society'). This is an entirely positive phenomenon, which has created a growing potential for peoples and their governments around the world to build strong and healthy science policies for their countries. Supporters and sceptics alike are becoming better able to hold effective discussions.

The growing trust and dialogue that are being created will allow, for example, open and honest progress to be made in pushing back the frontiers of genetic knowledge, while minimising the risks perceived by people who fear that 'tampering with nature' is the road to disaster.

In finding our way towards good policies, we will need to debate and discuss the relevance of the way science is carried out. The public interest can only be served if we optimize the structures we use in our different political systems to fund, organize, advance and control scientific progress. Whether or not individual members of society are interested in science, in the sense of being fascinated by it, is perhaps not as important as some in the scientific community have hitherto assumed.

[4] Eccleshall, R. and Walker, G. (1998) *Biographical Dictionary of British Prime Ministers*, p.236. Routledge, London.

But we do all have an interest in science, in the sense of having a stake in what is done in our names, or with our taxes. If we want to keep the activities of our scientists in line with our principles and expectations, we must have strong public debate about the ways in which our politicians seek to influence our scientific community through funding mechanisms, regulation and public policy.

This book is a contribution to such a debate, based around my own experience as a scientist and as a campaigner for science in the United Kingdom. In it, I have use the words 'science' and 'technology' in the general and overlapping senses that most of us use them in everyday speech, and have not chosen to be rigid in distinguishing between them. Research and its application are inextricably linked, each informing the other, so it does not seem necessary to demand that science and technology are always strictly delineated.

The examples I have chosen are mostly from my own country, but the principles they illustrate are by no means unique to the UK. They might apply equally to any country. In an increasingly globalized world, which is ever more influenced by science and technology, I doubt whether many countries will get through the next fifty years without experiencing some of the policy debates that are central to this book.

PETER COTGREAVE
January 2003

1 Paying our wages and the history of science

Examining the car-manufacturing industry demonstrates the need for a strong research base, while international comparisons and historical information show how wealth is generated from publicly-funded science

THE American President Ronald Reagan was known for his slips of the tongue. He once got into trouble for saying that he was going to bomb Russia, merely as something to utter when sound engineers asked him to test a microphone, and on another occasion, he called the British Princess Diana 'Princess David' for no obvious reason.

One of his mistakes that went largely unnoticed was at a Republican fund-raising event in 1982, when he said that he thought his administration was 'going to succeed' in 'trying to get unemployment to go up'. This was an easy mistake to make. In the course of natural speech, we probably all regularly say the opposite of what we mean, but the context of what we are saying, together with our tone, means that few people genuinely misunderstand us most of the time. Unfortunately, on this occasion, the President of the United States was not helped by the fact that his Budget Director, David Stockman, had said a few months earlier that 'none of us really understands what's going on with all these numbers'.[1]

In making these comments, Ronald Reagan and his aide were displaying two of the main preoccupations of any modern government. The first is that keeping the economy buoyant, so that relatively few people are out of work, is essential – Reagan knew that unemployment was important, even if he were unsure about which way it was supposed to go. The second preoccupation of politicians is that there is always another election to win, generally no more than four or five years away. Reagan knew that he had to convince people that unemployment would be reduced on a timescale of a few years at the longest.

And as his successor Bill Clinton famously said, the main issue on which elections are fought is undoubtedly 'the economy, stupid!' So,

[1] Dewdney, A.K. (1993) *200% of Nothing*, p.95. John Wiley & Sons, New York.

from a mass of information that is not only incomprehensible to most of us, but which even officials occasionally admit 'none of us really understands,' presidents and prime ministers must produce policies that keep people in jobs, and those policies must be seen to work on short timescales. Because they will have to face an election within a few years, they are unlikely to adopt policies that might be beneficial in the long-term, but which may seem harmful and unpopular in the short-term.

Scientific research is generally a long-term activity, and sometimes suffers from the short-term populism of the political process. And it is not only science itself that suffers from short-term thinking, but also many of the consequences of science, including all of the jobs that depend on the ability of research and development to keep a nation's economy ahead of the competition. Short-term actions, aimed at creating or preserving some jobs before the next political election, might actually harm the chances of creating more and better jobs in the longer term.

* * *

The issue of short-term thinking in relation to scientific research became of critical importance when the Rover motorcar company found itself in serious difficulties in the spring of the year 2000. The firm was owned by the German company BMW, but its main manufacturing base was at Longbridge in the city of Birmingham in the English Midlands. Almost seven thousand people were employed at the Longbridge site, with thousands of other local jobs indirectly dependent on the factory.

Rover was not performing well, and the firm's German bosses decided that they could no longer sustain the financial losses that the English company was suffering. They made the simple decision to sell the company as quickly as possible to whomever would take it off their hands – the eventual price was £10 sterling. The decision caused a huge fuss in British politics because its handling left the then Trade Secretary, Stephen Byers, with the unenviable task of scrambling to save thousands of jobs without, apparently, having accurate and detailed information on which to base his actions.

It was clear that whoever was to buy the Rover company, an enormous number of people would probably lose their jobs. For every job that was lost, a whole family would be plunged into financial, psychological and emotional difficulties. Bearing this in mind, nobody could deny that Rover's prospects really mattered, not just to thousands of individuals and their families, but to the nation as a whole, and in particular to a government that would soon be seeking re-election. The story dominated the news headlines for weeks, with reports of determined and urgent action by everyone from the Prime Minister downwards.

Depressingly, but not surprisingly, the issue of scientific research at Longbridge was barely mentioned. Everyone was focusing, quite understandably, on the immediate concern of saving as many jobs as possible within the few weeks that remained before the BMW team was threatening to pull the plug entirely and make everyone redundant.

The consortium that eventually bought Rover was called Phoenix. The bid made by Phoenix was supported by the British government, partly because the team was offering to retain the relatively high output of the factory on which so many jobs depended. Another bid, which would have safeguarded some jobs by drastically reducing the number of cars that were produced, but increasing their market value, was always destined to fail. It would have been politically unacceptable to overlook the possibility, however small, of retaining a larger number of jobs.

Desperate situations call for desperate measures and, in order to safeguard as many jobs as possible, the buyer had to slash the factory's costs. This meant that the research and development laboratories of the Rover company were doomed. There was said to be no way the new company could afford to invest in research, which employed relatively few people and which produced results that could not be made to turn a profit instantaneously.

In other words, the harsh realities of business, coupled with the political necessity and public pressure to save jobs, meant that another small section of British science had to be axed. About 600 workers were made redundant at the company's research and development facility. No doubt the company's buyers saw this as an acceptable sacrifice in their efforts to save the jobs of thousands of others. The local press, on the other hand, later said the researchers had been 'thrown on the dole queue' and the workers mounted a legal challenge to the decision to make them redundant.[2]

In the circumstances, the decision to endanger the future of research at Rover was not just unavoidable, it was right. We would be inhuman evangelists for science if we advocated saving such a laboratory at the expense of the jobs of thousands of young parents. However much society chooses to value science, we could not condemn the Rover workers' children to suffer the devastating financial and social consequences of suddenly finding their own parents on the dole, and all their friends' parents in the same situation. Parliamentarians from the left and right of British politics

[2] *Sunday Mercury*, 24 March 2001.

thought the deal was 'welcome'.[3] But just because a decision is right, it is not necessarily entirely desirable. The long-term future of any manufacturing industry depends on continually inventing new and better products. If one company does not do so, another one will, and when that happens, the first company's market will suddenly disappear.

The financial world disagreed with the government about the way forward for Rover. The press reported that the City of London had 'deep doubts' about the fact that the Phoenix consortium had not formed an alliance with a larger car manufacturer. Such involvement would bring 'crucial research and design skills', which were seen as 'crucial to the plant's long term future'.[4] Without some level of scientific and engineering investment, the 'money and expertise to research and develop the Rover models of the future will be lacking'. Despite the personal appeals of the Trade Secretary, 'Britain's banks [did] not want [the] Phoenix consortium as a customer'.[5]

* * *

The view that Rover needed science and engineering if it were to succeed was not idle speculation. There is compelling evidence that putting money into research and development is a sound investment for manufacturing industry. According to an analysis by the British government, companies involved in making pharmaceuticals, computer software and hardware, or electrical and electronic goods, all improve their performance by investing in research and development. There is a strong correlation between the amount of money that a company spends on research and development and the amount of goods and services that the company manages to sell.[6]

This is also true of the car manufacturing industry, where those companies that continued to invest in research during the recession of the early 1990s were the ones that were performing best by the end of the decade. Around the world, the two heaviest investors during the recession were both Japanese companies – in 1992, Toyota invested £21,860 per employee in research and development, while Honda invested £11,250 per employee. By 1999, the worldwide sales of cars by these two companies had grown by 21% and 37%, respectively.

[3] Angela Browning, the Conservative Shadow Trade Secretary said, 'We welcome the fact that the Phoenix bid has been secured', and Dr Lynne Jones, the outspoken left-wing MP for Birmingham Selly Oak spoke of 'today's welcome announcement'.

[4] *The Guardian*, 8 May 2000.

[5] *The Guardian*, 3 May 2000.

[6] *The 2000 R&D Scoreboard: Commentary and Analysis*, Department of Trade and Industry (2000).

The Italian company Fiat, which in 1992 invested only £3,890 per employee in research and development, had not grown at all by 1999. Pirelli, another Italian firm, whose research investment was just £2,070 per employee, has performed disastrously, and has seen its sales shrink by 28%. Across the industry, the trend for good performance to follow investment in research was extremely strong. Nobody could possibly question that the companies that invested most heavily in research were those that went on to achieve the best financial returns, and to create the most jobs. Between 1992 and 1999, the combined workforces of Honda and Toyota grew by 97,000, while Pirelli and Fiat shed a total of 90,000 jobs.[7]

The lesson for Rover at Longbridge is clear: do not give up on research just because times are hard at the moment. Nobody would pretend that research is the only driver of success in the car industry, which is an immensely complex business.[8] Experts agree that good advertising, strong supply chains and many other factors are crucial parts of a successful car manufacturing business.[9]

But science and engineering research are also crucial parts of success, and unless it rebuilds its research capacity in time, Rover will not survive. If the laboratories remain permanently closed, and if the company does not find some other source of high-quality research, it will always be several steps behind the competition. Disaster will not have been cancelled; it will merely have been postponed. It will be only a matter of time before the communities of Longbridge experience the hell of large-scale unemployment. Yet astonishingly, when the UK Parliament discussed the future of Rover on the day that Phoenix finally bought the company, there was no mention of research, either by the minister responsible or by any of the parliamentarians who questioned him on the deal.[10]

* * *

Let us assume that the weight of evidence prevails on Rover's new owners, and that, as time passes, the company manages to build up new research capacity, so that exciting new cars are created and sold, securing thousands of jobs in Birmingham and the surrounding area. In the meantime, somebody must keep the fire burning.

[7] Data taken from *The 1993 UK R&D Scoreboard: Company Reporting*, Company Reporting Ltd (1993) and *The 2000 R&D Scoreboard: Company Data*, Department of Trade and Industry (2000). Growth figures are adjusted to real terms on the assumption of 2.5% inflation per year.

[8] *Motor Industry of Great Britain 1998: World Automotive Statistics*, Society of Motor Manufacturers and Traders, 1998.

[9] Rhys, G.R. *et al.* (1993) *The London Motor Conference.* Financial Times, London.

[10] *Hansard* (House of Commons) 9 May 2000, columns 645–657.

When the company is back on its feet, Rover will not be able to conjure a team of scientists and a working laboratory out of thin air. Apart from anything else, it will need to find people who know something about the subjects that matter. It may be able to poach a few researchers from other car companies, but those other firms are not in the business of training hordes of staff who will leave at the drop of a hat. They will wish to keep their best researchers and since they will be successful companies, they will be able to offer those researchers salaries and other inducements that will make many of them reluctant to move to a firm with a shaky reputation.

The fact is that, in a successful economy, it is the job of the government to ensure that the country has the capacity to react to new opportunities. When those opportunities happen to include the possibility of an economic gain through technological progress, then an advanced country should be in a position to try and take advantage. In many cases, such a strategy will mean maintaining a scientific capacity that remains unused for years, capable of springing into action and flourishing, without warning.

There are those who believe that, as the economy becomes more and more based on tourism and the service sector, it is less important to worry about manufacturing, agriculture and other science-based industries. But their arguments ultimate seem absurd – we cannot all make our living by renting out holiday homes and selling one another cappuccinos. Successful economies must generate new things that other people are prepared to pay for.

It is barely possible to predict what scientific capacity will be needed to achieve this constant innovation, so the only sensible strategy is to maintain as broad a science base as possible, and to keep abreast of what is happening in other countries. Short-term thinking, in which we choose to focus only on those things that may be useful within a few years, can be a disaster for science, because it is simply not possible to produce breakthroughs on a timescale determined by the electoral cycle. If Alexander Fleming had been forced to produce results within four years, it is extremely doubtful whether he would have discovered penicillin.

It is impossible to overstate the importance to the economy of a broad base of fundamental scientific inquiry, and the best technologists have always appreciated that science is the basis of much of their work. Perhaps the most prolific technological innovator who has yet lived was Thomas Edison, who held patents on a record 1093 innovations, and who is best known for useful inventions such as the first sound-recording equipment and the first commercial electric lighting system. He knew that his advances would not have been possible without basic scientific research.

When Edison's telegraphy tests for the British Post Office were only partly successful in 1874, he realized that there was a great deal he did not know about the electrical and chemical phenomena involved in cable telegraphy. He began to equip a complete electrical and chemical laboratory at his shop in Newark, New Jersey. From then on, he fully understood the value of scientific investigation, and began to boast that his laboratory contained 'every conceivable variety of Electrical Apparatus, and any quantity of Chemicals for experimentation'.[11] His later inventions, developed over a period of 57 years, were all grounded in an experimental approach and an ever-increasing base of scientific knowledge.

One of the clearest modern examples of how basic research can lead to economically useful products comes from the drugs known as monoclonal antibodies. The problem of how to produce large quantities of a single, known antibody, without contamination from other molecules, was solved in 1975 by Georges Köhler and César Milstein, working in Cambridge. They published a major and definitive paper on the subject in one of the world's leading scientific journals, *Nature*,[12] and were later awarded a Nobel Prize for their work.

Their work was funded by the UK's Medical Research Council because it was interesting, not because anybody saw any practical value in it, and Milstein later spoke about having 'been attracted to the puzzle' and of the 'fundamental emphasis', rather than practical application of the work.[13]

Soon after this breakthrough, people began to realize that monoclonal antibodies might be useful in the treatment of various medical conditions, and today, almost all of us know someone who has an illness against which these drugs are used. Breast cancer, leukaemia, rheumatoid arthritis and heart disease are all treated with monoclonal antibodies, and the drugs are also used in combating the problems associated with organ rejection following kidney transplants.

Monoclonal antibodies, which sprang from Köhler and Milstein's attraction to a fundamental puzzle, are now the basis of an industry worth billions of dollars a year. A single drug, one of thousands now on the market, generated sales worth US$365 million in 1998.[14] Known as abciximab, it is used to prevent blood-clotting in patients with heart disease,

[11] Quoted by Israel, P. (1998) *Edison: A Life of Invention*, pp.85–86. John Wiley & Sons, London.

[12] Köhler, G. and Milstein, C. (1975) Continuous cultures of fused cells secreting antibody of predefined specificity, *Nature*, Volume 256, pp.495–497.

[13] Milstein, C. (1981) 12th Sir Hans Krebs Lecture: From Antibody Diversity to Monoclonal Antibodies, *European Journal of Biochemistry*, Volume 118, pp.429–436.

[14] Centocor Inc. *Annual Report*, 1998.

and the American company that makes it, Centocor Inc, was bought in 1999 by the multinational pharmaceutical giant Johnson & Johnson in a deal worth US$4.9 billion. The main reason for the purchase was that Centocor was 'a leader in monoclonal antibody technology'.[15]

It is impossible to calculate the scale of the wealth that is now created (let alone the health benefits that are generated) by monoclonal antibodies. With new drugs being developed all the time, it is certainly impossible to predict their future potential. What is not in doubt is the huge economic return of the initial investment. The Medical Research Council, the body that funded the original work, has a total annual budget that is roughly equal to the yearly sales of abciximab alone.[16]

* * *

Every so often, when there is not much real news to report, the media will 'expose' a story about how university researchers are carrying out some crackpot study that is portrayed as being manifestly a ridiculous waste of public money. In 1999, Len Fisher from the University of Bristol was castigated for calculating a scientific formula for the best way to dunk a biscuit into a hot cup of tea or coffee. Apparently, the 'flat-on' approach, in which the biscuit is dunked face down is superior to the 'edge-on' method, because it prevents the biscuit from breaking in half.[17]

In a round-up of what he considered to be the most pointless research of 2001, David Derbyshire noted that scientists had discovered that ducklings 'display evidence of fear' when you show them photographs of foxes, that victims of school bullying were 'more worried about going to school than the bullies', and that 'some college students are performing poorly because they spend too much time, especially late night hours, on the internet'.[18]

Stories like these create an opportunity for some people to criticize the way in which our taxes are used to pay for pointless science and no doubt on some occasions, they have a point. Since scientists are only human, a small number of them will be lazy, silly or irresponsible. There is even evidence that some are dishonest, although their peers can be ruthless in exposing scientific fraud.[19] More often the media hype surrounding a

[15] Johnson & Johnson *Annual Report*, 1999.

[16] *SET Statistics 2000: A Handbook of Science, Engineering and Technology Indicators*, p.5. Department of Trade and Industry/Office of Science and Technology, London [Cm 4902].

[17] *The Times*, 7 June 2001.

[18] *Daily Telegraph*, 2 January 2002.

[19] Grayson, L. (1995) *Scientific Deception: An Overview and Guide to the Literature of Misconduct and Fraud in Scientific Research*. British Library, London; Grayson, L. (1997) *Scientific Deception: An Update*. British Library, London.

story hides the fact that there is an underlying grain of respectability and scientific integrity in the research that is being reported.

More serious than media reports of 'useless' science are those people who think that science is interesting and important, but who feel nevertheless that it should not be the role of the state to fund scientific research. According to this view, private capital is the proper source of funding for research because, in the end, it is private industry that will reap the financial rewards. Such free-marketeers are generally critical of the increase in the number of areas of life into which the government seeks to poke its nose, and think that science is just one more area where politicians ought to be banned from interfering.

There are many flaws in this argument, but its fundamental position contains a kernel of truth. Scientists often back up their arguments for increased funding by pointing out the economic benefits that have been derived from science in the past. It is true that private enterprise is generally one of the ultimate financial beneficiaries of good research as the examples of Honda, Toyota, Centocor and Johnson & Johnson demonstrate. Moreover, the public purse can never afford everything that taxpayers want of it, and schools and hospitals are always in equally dire need of investment as are research laboratories.

The greatest exponent of the idea that there ought not to be any public funding for research is Dr Terence Kealey, who was a biochemist at the University of Cambridge before becoming Vice Chancellor of the University of Buckingham.[20] He is considered eccentric by many who campaign for science and technology, but there is no question that he provides them with a useful service, because he is constantly appearing in the science and business media, and his pronouncements usually provide an opportunity for the rest of us to put the opposite case; the media, after all, love an argument.

Kealey's principal argument is that public funding for research merely displaces private funding.[21] He subscribes to Ronald Reagan's opinion that a taxpayer is 'someone who works for the . . . government but doesn't have to take a Civil Service examination'.[22] Consequently, he wishes to cut out any public expenditure that could be better paid for by the private sector. Industrialists would fund more scientific research, he believes, if it were necessary, but they can see no reason to spend their own money when the bloated public coffers seem to be only too willing to pay up.

[20] The arguments are put strongly in Kealey, T. (1996) *The Economic Laws of Scientific Research*, Macmillan Press, London.

[21] ibid.

[22] Jay, A. (1999) *Oxford Dictionary of Political Quotations*, p.301. Oxford University Press, Oxford.

The flaws in such arguments are fundamental to the question of why science really should be of concern to non-scientists, going about their daily routines, minding their own business. If it is true that private companies would, if required to do so, fund all the research that is needed by society, then the public interest would certainly be served by slashing government funding for science. The United States government alone spends about US$18 billion each year on research and development.[23]

Perhaps the simplest argument against Kealey is one with which he probably would not disagree, and it raises an ethical issue. When unexpected problems arise, governments need sources of expert advice and information and, insofar as it is possible, that information must be impartial. A typical example would be issues of safety, where action may be needed rapidly. When the Chernobyl nuclear reactor exploded, and radioactive material spread across Europe, governments had to decide what to do about it. Likewise, action was needed when the American government realized that there might be some reality in the condition known as Gulf War Syndrome, in which veterans of the Middle East campaign of 1991 experienced a variety of unexplained symptoms.

Whatever kind of society we live in – whether it is democratic or undemocratic, fascist or communist, socialist or capitalist – somebody has to make the choices that govern us. How high should taxes be? What are the boundaries of legal and illegal behaviour? Are some activities too dangerous to be permitted even as a matter of personal choice?

Assuming that all of these choices are made with the aim of maximizing or optimizing some measure of human happiness, then they need to be made on the basis of good information. The kind of happiness that governments seek to create for their peoples can be highly variable, but the basic objectives tend to be things like good health. Other basic objectives, such as wealth, education, and freedom contribute to happiness, and in civilized countries, happiness is believed to be enhanced by access to the arts, music, science, and other activities that allow people to express their individual creativity.

The point about these decisions is that, if they are to be of any value to anyone at all, they can only be made on the basis of good information. And since decisions are supposed to be taken for the general good of all the people, it is dangerous to obtain that information exclusively from anyone with a highly personal interest in the decision, especially if that interest appears to diverge from the general good. Put simply, it is danger-

[23] *Main Science and Technology Indicators*, Organisation for Economic Co-operation and Development.

ous to ask big business to advise you on the safety of products that the same big business wants to sell.

That is not to say that scientists working in the public sector should be treated with blind deference, as if they are saints by comparison with industrial researchers. They are human, and have human failings, and in situations where they are called upon to advise Government, they are often forced to offer opinions rapidly, and to base them on incomplete information. They may have to extrapolate the results of a small laboratory-based experiment to a much larger and more complex real-world situation, or even simply to offer a 'best guess'. For these reasons, publicly-funded scientific advice to public authorities should be closely scrutinized, but there is no question of its necessity. A publicly-funded science base is essential to good democracy.

* * *

The ethical arguments for the need for publicly-funded research are undeniable and in truth, no other argument is needed. But the fact is that even without recourse to ethics, the economic arguments are strongly in favour of taxpayers supporting science. Comparing the industrialized nations of the world reveals an interesting correlation between the degree to which public policy supports scientific endeavour and the degree to which private companies choose to invest in science.

The country whose government invests most in scientific research, scaled for the size of the population, is the USA, and that is also the country where private businesses invest most in research. France also has high investment in research from both the government and the private sector. The two countries in the Group of Seven industrialized nations whose governments invest the least in research per head of the population are Italy and Canada, and it is in these countries that business investment in research is lowest. The UK has the third lowest government support and also the third lowest private support. This pattern cannot have arisen simply because large countries have more money overall, because the analysis ensures that levels of investment are adjusted for the size of each country's population.

A correlation between government support for science and business investment in science does not necessarily imply that one causes the other, but it seems likely that private enterprise is following the lead of governments. Where governments support science, businesses have access to strong public sector collaborators and a good flow of skilled workers who have been trained in vibrant laboratories in universities and colleges. Consequently, private industry sees great advantages in building and running its own laboratories.

The case is further strengthened by examining how investment in research changed during the 1990s, which many countries started by experiencing an economic recession. The governments of most industrialized countries reduced their investment in research between 1988 and 1998, as tax receipts fell and, in most countries, the private sector also contracted its research base. But the countries in which government cuts were greatest tended to be those in which private investment was also hardest hit. The only developed countries in which public support for research rose during the 1990s were Japan and Finland. These two countries were also the only ones in which businesses held their nerve, and in which private sector investment either rose slightly or suffered a very marginal reduction.[24] As the example of the motorcar companies makes clear, businesses that hold their nerve in hard times are the ones that benefit when economic conditions pick up.

It seems that Gary S. Becker, the Nobel Prizewinning economist, was correct when he wrote that in the USA 'higher federal funding of basic research . . . usually increases rather than decreases spending by . . . companies'. He was focusing on the pharmaceutical industry, although his arguments might equally apply to any kind of business where innovation is important. He went so far as to say that this correlation was the reason 'why economists agree that governments should support basic research'.[25]

* * *

History provides the final nail in the coffin of the idea that public money is wasted on research. The arguments of those who oppose government spending on science rely in part on historical data which purport to show that the great scientific and technological progress of the past was largely financed through private investment, without the wasteful dead hand of government.

Terence Kealey, for example, argues that both the industrial revolution and the agricultural revolution in the UK were financed by private individuals and partnerships, and not out of the public purse. His illustrative examples include J.B. Lawes, who discovered the fertilizer superphosphate. As a 'wealthy farmer', Kealey argues, Lawes was able to 'turn his

[24] Data are taken from the *Main Science and Technology Indicators* and *Basic Science and Technology Indicators* of the Organisation for Economic Co-operation and Development, and the analysis is presented in *Science Policies for the Next Government: Agenda for the Next Five Years*, pp.39–40. Save British Science Society, London (2001).

[25] Becker, G.S. (1999) Save some of the surplus for medical research, *Business Week*, 19 April 1999, p.27.

farm at Rothamsted into a laboratory', and he ridicules the idea that people like Lawes should be 'taxed so that agricultural experiments could then be funded by the State'. Kealey also points out that the Rothamsted Experimental Station is still producing 'wonderful research'. He fails to observe that the Station is now largely funded by the State, or that research careers are no longer restricted to the rich.

One possible argument against the use of the agricultural and industrial revolutions as comparisons with modern science is that they were based on rather low-technology industries, such as cereal production and basic manufacturing of cloth. It is true that an eighteenth or nineteenth century farm or factory would seem primitive to modern eyes, but at the time, the new research and technologies really were revolutionary; they allowed their users to create the modern age. The Luddites smashed farm machinery because rapidly developing technology was having an enormous effect on the lives of many people.

The real problem with the argument that historical scientific success was built on private funding is that it ignores the complexity of public financing of research in previous centuries. The details of much of that funding are utterly impenetrable and in the UK are largely hidden among millions of documents in the Public Record Office.

* * *

A proper survey of public funding for scientific research and development would be a huge project, and has never been undertaken for any historical period before the twentieth century. However, a few examples have surfaced as part of other historical investigations and deserve a little attention because they shed light on the vast array of public sources that have funded research and development, and on the enormous diversity of scientific and technological endeavours that have been supported.

Terence Kealey devotes a chapter of his book to the agricultural revolution. He discusses the research and development that was involved, but also acknowledges the importance of the enclosure of common land into workable fields, specifically mentioning the 'wealth creation that the enclosures accelerated'.[26] He does not, however, explore the ways in which the enclosure of agricultural land was financed.

In fact, there were considerable public costs of enclosure in England and Wales, levied as a local rate to cover the costs of surveyors and administrators. These amounted to as much as £3 per acre, and about eight

[26] Kealey, T. (1996) *The Economic Laws of Scientific Research*, p.59. Macmillan Press, London.

million acres were enclosed in England during the half-century following 1760.[27] The total public costs of this period of enclosure were thus in the region of £24 million in England alone, which is roughly equivalent to six months' worth of England's national output.[28]

The equivalent tax burden today could be estimated at £28 per month from every taxpayer, an increase in the basic rate of income tax of more than a penny in the pound.[29] In other words, it is simply nonsense to suggest that the wealth-creating potential generated by the technological advance of the agricultural revolution was funded without a substantial input of public money.

The more one looks, the more astonishing it becomes how many different mechanisms have existed historically for the support of science and technology in the UK. Indeed, support for scholarship in the broadest sense dates back at least as far as the late ninth century when King Alfred the Great gave state support for schools and believed that 'punishments befell us when we did not cherish learning'.[30] In the sixteenth and seventeenth centuries, Francis Bacon championed state support for science[31] and is generally regarded as the earliest strong proponent for research in something like its modern form.[32]

In the eighteenth century, the formation of the Board of Longitude created perhaps the earliest forerunner of modern Research Councils. It dispensed over £100,000 between 1714 and 1828, which is the equivalent of roughly £250,000 per year in the twenty-first century. Like a Research Council, a panel of experts made judgements about the best scientific advances to support with public money. Like a Research Council, it dispersed its funds among a variety of sources and supported work that ultimately failed in its aims as well as work that succeeded. It differed from modern funding bodies in that most of its grants were given after the research had been carried out rather than before, and in concentrating on a single problem – how to make an accurate determination of longitude at

[27] Rule, J. (1992) *The Vital Century: England's Developing Economy 1714–1815*, p.86. Longman, London. Up to one quarter of England's 32 million acres are estimated to have been enclosed in two bursts around 1775 and 1805.

[28] Holmes, G. and Szechi, D. (1993) *The Age of Oligarchy: Pre-industrial Britain 1722–1783*, p.382. Longman, London. The national product of England and Wales was about £70 million in 1760.

[29] According to the Organisation of Economic Co-operation and Development, the Gross National Product of the UK is roughly £1 trillion. Divided among the active workforce of roughly 30 million people and spread over half a century, one half of this is £333 per year.

[30] *Alfred the Great: Asser's Life of Alfred and Other Contemporary Sources*. Penguin, London (1983).

[31] Bacon, F. (1925) *The Essays, Colours of Good and Evil, and Advancement of Learning*. Macmillan, London.

[32] Jardine, L. and Stewart, A. (1998) *Hostage to Fortune: The Troubled Life of Francis Bacon 1561–1626*. Phoenix Grant, London.

sea – rather than a broader spectrum of scientific inquiry. But in focusing on solving a problem that would improve wealth creation and quality of life in the future, it was identical to that part of modern Research Council funding that is targeted towards the priorities identified by Foresight Panels, which attempt to second-guess the future technological needs of society and the economy.[33]

Parliamentary grants were a favourite form of public support for scientific and technological invention in the early nineteenth century. Examples include Edmund Cartwright, the inventor of the power loom, who was granted £10,000 in 1809, and John Loudon McAdam, whose experiments led to the invention of the 'macadamized' road surface, and who was voted £2,000 in 1825.[34]

State support also came in the form of royal patronage, which sometimes involved real money. For example, King William IV granted a lifetime's pension of £100 a year to William Smith,[35] the 'father of geology', and Prince Albert was well known as the major driving force behind the Great Exhibition of 1851 which sought to bring together much of the world's greatest science and technology, and the proceeds of which are still funding research fellowships today.[36]

Later in the century, the palaeontologist Richard Owen found favour with the establishment and, in addition to receiving a royal stipend, was given a rent-free house by Queen Victoria. Her Secretary, Mr Phipps, told Owen that 'the Queen commands me to say that she thinks that there is no method in which she can better give a tribute of her respect and regard for science'.[37]

The first half of the twentieth century saw a new way of rewarding scientific and technological advance, particularly in relation to inventions that were, or might have been, of use in wartime. The first Royal Commission on Awards to Inventors ran from 1919 to 1937 and dealt with more than 700 inventions, including microphones and telegraphic devices.[38] In 1921, the Coslett Syndicate was awarded £300 for inventing a process to prevent iron and steel from rusting.[39] A second Commission followed World War Two.

* * *

33 Sobel, D. (1996) *Longitude*. Fourth Estate, London.

34 Magnusson, M. (ed.) (1990) *Chambers Biographical Dictionary*, pp.269 and 931. Chambers, Edinburgh.

35 Winchester, S. (2001) *The Map that Changed the World*, p.293. Viking, London.

36 www.royalcommission1851.org.uk.

37 Cadbury, D. (2000) *The Dinosaur Hunters: A True Story of Scientific Rivalry and the Discovery of the Prehistoric World*, p.288. Fourth Estate, London.

38 Public Record Office: Piece List for Class T173.

39 Public Record Office: T173/177.

Some people think that scientists will always argue for more public funds. According to the writer Stephen Shapin, 'scientists' capacity to absorb money expands in relation to the funds available'.[40] There is some truth in Shapin's assertion, because there is an infinite universe to understand, and each new discovery opens up new lines of enquiry. In effect, 'the main point of answering one question is merely to work out what the next one is'.[41] Any government has many calls on its funds, and science must always take its place in the queue along with healthcare, education, defence and industrial policy.

But since science contributes to each of these areas of public policy, it must occupy a special place and must be treated, at least in part, as a public good. It is neither ethically desirable nor particularly successful to expect the private sector to carry out research of the kind that governments are best placed to perform. Nor is it true that, historically, governments did not interest themselves with the funding of scientific research and development, on the grounds that it was the exclusive preserve of private capital to take the risk of funding potentially profitable developments. The best way for governments to encourage business to conduct research and development is to show a lead, and carry out a high level of publicly funded research.

For society to benefit fully from science, it is equally important that governments organize research in ways that promote the public good and that citizens, voters and private industry use their voices to encourage the best ways of funding and arranging the nation's scientific endeavours. The ramifications are important for the politics of the environment, education, healthcare, defence and retirement pensions. Indeed, they affect every aspect of public policy, and every aspect of life.

[40] Shapin, S. (2001) Guests in the President's House. *London Review of Books*, 18 October 2001, p.6.

[41] Cotgreave, P. (2001) Christmas Cheer: The Dream Science Base, *The Biochemist*, December 2001, pp.7–9.

2 Railways, the internet and old age

Comparing the current information revolution with similar developments in history suggests that economic sustainability depends on investment in research-based companies, not just those that use high-technology

In Alistair Beaton's comic play, *Feelgood*, several speechwriters are preparing the text from which a fictional Prime Minister will address his Party's annual conference. The play is a satire on British politics and the fictional Party leader is very clearly based on the UK's Labour Premier, Tony Blair. In an attempt to make him appear a man of the people, one of the aides suggests that the Prime Minister should list the 'ordinary' things he worries about – his children's future, the environment, his mortgage, and so on.

At this point, the other speechwriter treats his colleague as a fool, because everyone knows that the Prime Minister does not have to worry about a mortgage. Just as the American President lives in the White House for his term of office, or the French Prime Minister lives at the Hotel Matignon in Paris, the British Prime Minister lives rent free in Number 10 Downing Street. After all, he or she needs a flat in Westminster, and he or she can hardly spend the first few weeks in office looking for a place to live when they should be running the country.

There is, however, perhaps one 'ordinary' issue that senior politicians do need to worry about – their pensions. It may not be true, as Enoch Powell once said, that 'all political careers . . . end in failure,'[1] but nobody stays in office forever and everyone needs a source of income in their retirement. Many of the most senior political leaders write their memoirs or give lecture tours when they retire but, like everyone else, they must save some of their income for the future.

Beaton's play stresses the cynicism of modern politics. If politicians were genuinely cynical about pensions, they would rarely worry about them as a political issue. As long as there is likely to be enough money in the coffers to pay the state's liabilities for the next few years, they do not

[1] Jay. A (1999) *Oxford Dictionary of Political Quotations*, p.297. Oxford University Pres, Oxford.

need to worry because they will be out of office by the time problems arise. Their successors will be responsible for sorting out any problems.

But because their own pensions are, by definition, something over which they will have no political control when the time comes, they need to ensure that long-term policies are in place to safeguard their own investments. Perhaps that is one reason why pensions are traditionally a major political issue. After the 2001 General Election, Tony Blair even reorganized the government's administration and created a Department of Work and Pensions. He was following in a long tradition of treating retirement pensions as an extremely important political issue.

Indeed, when old-age pensions were introduced in the UK in 1908, they caused one of the greatest upheavals ever in the nation's politics. The Liberal Party finance minister, David Lloyd George, was loved by the population when he brought in the first scheme, but he 'merely accentuated the [existing] crisis in public finance'. Always combative, Lloyd George had to propose a hike in taxation but, in doing so, he introduced the idea of pensions into his official budget, which was the sole preserve of the House of Commons; the Lords had no authority on the budget. These tactics were contrary to the constitutional convention that the House of Lords could vote on everything except strictly economic matters. The constitutional crisis that ensued saw the introduction of the Parliament Act, which for the first time ever gave the elected Commons almost absolute power over the hereditary Lords.[2] Arguably, this political bloodshed contributed to the downfall of the Liberal Party and the rise of the Labour Party. A century later, the Labour Party leader Tony Blair was still dealing both with expensive pension policies and with unfinished constitutional reform.

* * *

Pensions come in two forms – 'funded' schemes and 'pay-as-you-go' schemes. In funded schemes, each of us builds up our own pot of money, which is invested (on the stock market or elsewhere) and grows throughout our working life, and from which we can draw an income in our retirement.

Pay-as-you-go schemes, like the one that funds the British state pension, is rather like a 'pyramid' investment scheme – those of us who are working today pay taxes and National Insurance contributions that are used immediately to pay for the pensions of our own parents and grand-

[2] Ramsden, J. (1998) *An Appetite for Power: A History of the Conservative Party since 1830*, pp.207–208. HarperCollins, London.

parents. This is not unlike the situation that has pertained through most of history, in which families were responsible for looking after their own elderly members. The big difference is that, with a state pension scheme, risk is spread among us all so that we will continue to have an income in our old age, even if we have no families, or if our children happen to be poor, useless, mean, cruel or dead.

We now recognize that the big problem with such pay-as-you-go pension schemes is that the politicians who make promises now, about our pensions in the future, actually have no knowledge whatsoever of how their successors will pay for their promises; they have no idea how many people will be paying how much tax in twenty or thirty years' time.

This kind of scheme has worked in industrialized countries in the past century largely because, on the whole, national economies have performed well, and because most people have spent forty or fifty years paying taxes and only ten or twenty years drawing a pension. Indeed, when Bismarck introduced the first modern state-funded old-age pensions in Germany in the late nineteenth century, it is said that he set the retirement age at 65 years because he estimated that most people would end up drawing a pension for just one year after paying taxes for about fifty years.

In the modern world, most of us spend fewer years working and then live much longer. In an age where almost half of all young people study at university, it is common for people not to join the labour force until they are in their early twenties, nor is it remarkable for us to retire in our fifties. Since many of us now live well into our eighties and nineties, a large number of people are spending 30 or 35 years making contributions to their pension schemes, and then expecting payouts for an equally long period.

Not surprisingly, governments around the world are moving away from state provision and towards funded schemes. To secure decent pensions into the future, they expect us to make our own provision now. The British minister with responsibility in this area, Jeff Rooker, said in January 2001 that he wanted 'to ensure that as many people as possible retire with a decent second pension on top of their basic state pension',[3] while the British Chancellor of the Exchequer, Gordon Brown, in claiming to have increased 'financial independence', wanted to 'go further and encourage people to make provision for financial security throughout their lives'.[4]

[3] *Hansard* [House of Commons] 18 January 2001, column 387W.

[4] *Prudent for a Purpose: Working for a Stronger and Fairer Britain: Economic and Fiscal Strategy Report and Financial Statement and Budget Report March 2000*, p.94. The Stationery Office, London (2000). [House of Commons No. 346].

When Tony Blair reorganized the UK structure of government in the spring of 2001, he demonstrated his belief in the link between personal provision while we are in work and our incomes in retirement, by creating a Department of Work and Pensions. Pensions policy had previously been dealt with by the Department of Social Security, with the connotation that pensions were a benefit provided socially.

* * *

Funded pensions work because the money we invest when we are young grows, as those who manage our pension funds buy shares in successful and profitable companies. Politicians tell us that privately funded pensions will work for us now because the stock market will do well in the 'knowledge economy', so the money each of us hands over to a pension provider will grow and provide a healthy nest-egg for an active, healthy and happy retirement.

Gordon Brown claimed that people on modest incomes would be able to make pension contributions because the government was 'committed to creating a stable economy capable of delivering sustained growth' and that it was achieving its aims by ensuring that 'the UK is equipped to respond to the new opportunities provided by rapid globalization, the growth of new technologies, associated with the internet and electronic commerce in particular'.[5]

Mr Brown's claim was based on the very obvious high performance of many high-technology companies. For example, in March 2000, the London Stock Market saw a 'defining moment', when nine biotechnology, computing and telecommunication companies joined the *Financial Times* 100 Share Index.[6] Reflecting a boom in high-technology shares, companies like Psion which manufactured palmtop computers, and Celltech, the UK's premier biotechnology group, joined the flagship index[7] which includes only the largest and most successful companies, typically those worth at least £3 billion.[8] Psion and Infobank 'took advantage of their rocketing share prices . . . to fund business development'.[9]

Kerrin Rosenberg of the actuarial company Bacon and Woodrow said that pension funds would need to start investing more heavily in internet

[5] *Prudent for a Purpose: Working for a Stronger and Fairer Britain: Economic and Fiscal Strategy Report and Financial Statement and Budget Report March 2000*, pp.36 and 94. The Stationery Office, London (2000). [House of Commons No. 346.]

[6] *The Guardian*, 8 March 2000.

[7] *The Times*, 20 March 2000.

[8] *The Times*, 4 September, 2000.

[9] *The Independent*, 3 March, 2000.

and other high-technology companies, because many 'had been caught with an underweight position in the technology sector'.[10]

However, the assertion that, with a stable economy, we could all have confidence in a knowledge-driven stock market to nurture our pension funds has turned out to be far too simplistic. History demonstrates that there have been many previous 'information revolutions' and that those who have invested in the new enterprises formed to exploit them have not always succeeded. Success has come not just from using new technology, but also from carrying out the research and development needed to invent wholly novel products and new ways of working.

* * *

It is fashionable to talk about an 'information revolution'. Some people use the expression to mean that the one-off introduction of the worldwide web has made information more readily available than it was before, while others are more concerned with the increasing speed with which it is possible to transfer information, by email or other electronic means.

Whatever method we choose to define the current information revolution, its basis is found in the fact that the speed at which computers can process information doubles roughly every 18 months. This rule of thumb, which is known as Moore's Law, has held for many years, and is part of the reason that home computers are both cheaper and more powerful than they were even a few years ago. Many scientists and industrial analysts believe that even as we approach the limits of conventional computing technology, the rule will continue to apply, because we will be able to build 'quantum computers', in which we will be able to use a single atom to store each unit of information.[11] Even if these optimists turn out to be wrong, the fact that computing power has increased steadily for decades means that we have some claim to live in the 'information age'.

However, the current 'information revolution' is by no means the first. The Romans revolutionized communications technology in Europe, North Africa and the Middle East by using a series of beacons on hilltops, and by building good quality roads, which allowed improved travel for people, goods and information. This was the first in a long line of technological and other developments that significantly increased the speed at which information has been transmitted. One of the most significant improvements in the speed at which information could travel around Britain was the revolution in the building and financing of road transport that occurred in the eighteenth century.

[10] *The Times*, 26 February 2000.

[11] Blatt, R. (2001) Delicate information. *Nature*, Vol. 412, p.773.

After the Romans had left Britain, highway maintenance became an inefficient process which relied on the inhabitants of towns and villages to carry out their fair share, based on the amount of land they occupied. The work was hard, involving quarrying and carrying flagstones. If a villager refused to do his duty, either the work had to be done by others, or it would wait while the village constables encouraged the magistrates to coerce the defaulter to do his duty. Thousands of people must have found themselves in the same situation as the Cheshire man who, in 1623, was prosecuted because he 'neither wrought nor brought stones' and was 'behind with the carriage of sand and stones for the Highway'.[12] Each parish in England had to elect unpaid Surveyors of the Highways to assess what needed to be done, to encourage their neighbours to do it as best they could, and to prosecute those who refused to help.

But in the eighteenth century, improved road technology and financing brought noticeable benefits. Between the 1670s and the 1740s, communications improved significantly, partly because of the development of good postal services, and the eighteenth century road improvements added to the speed at which information could travel.[13] Roads that had been impassable in winter were usable all year, and travel became much faster. The journey from London to Shrewsbury had taken four days in 1753, but by 1772 it took only a day and a half. By 1815, the same trip lasted just 12 hours.[14] Just as Moore's Law describes how computing speeds have increased steadily for decades, so stagecoach timetables described how the transfer of information across England was increasing steadily in the 1700s.

And, just as many people now believe that the current 'information age' will help generate the economic growth needed to generate decent pension funds, so the eighteenth century advances were thought to be significant drivers for strengthening the economy, and for generating jobs and wealth. Constant improvements in the velocity at which people, goods and information travelled were financially valuable. Daniel Defoe, better known for writing *Robinson Crusoe*, sounded exactly like a twenty-first century internet entrepreneur when he said in the 1720s that 'as for trade, it will be encouraged by it every way; for carriage of . . . goods will be much easier . . . which will bring goods cheaper to market'.[15]

[12] Cheshire Record Office: QJF 52/2/Trinity 1623. folio 33.

[13] Steele, I. (1986) *The English Atlantic 1675–1740: An Exploration of Communication and Community*, p.213. Oxford University Press, New York.

[14] Royle, E. (1987) *Modern Britain: A Social History 1750–1985*, p.9. Edward Arnold, London.

[15] Defoe, D. (1962) *A Tour through the Whole Island of Great Britain*, Vol. 2, p.132. Everyman edition, London.

But even the new roads seemed backward when the age of steam trains began. The railway age is traditionally dated from 1830, when a passenger line opened, running trains between Liverpool and Manchester. Eventually, this entirely new form of travel completely revolutionized the ease and speed with which both people and goods could be transported. Later the electric telegraph made an even greater contribution to the speed at which information flew around the globe. By the end of the nineteenth century, communications were unimaginably faster than they had been a few decades earlier.

In 1820, the fastest communication between Bombay and London took 121 days. By 1840, this had been reduced to 43 days and by 1870, the telegraph system allowed London newspapers to publish news about what had happened in Bombay the previous day.[16] One commentator later thought this was evidence of the 'first information superhighway', observing that in some areas, the fastest postal service in the nineteenth century was 'as quick, if not quicker, than the "high tech" service we endure today'.[17]

Wireless transmission brought about yet another information revolution, perhaps the most similar to the internet boom. After Marconi transmitted the first wireless transatlantic message in 1901, the development of domestic radio receivers was rapid in the early decades of the twentieth century. Like the internet, radio revolutionized communications over vast distances, and over time, the market for domestic equipment became huge; most of us now own several radios.

So the advent of the internet, and of electronic business, is just the latest in a series of continual revolutions in the speed at which information travels. Just as Defoe believed the new roads were a staggering development, physicists now tell us that they are 'experimenting with a new type of computing based on quantum effects that could completely revolutionize information technology'.[18] Moreover, there are remarkable similarities between modern internet companies and some of the groups that drove earlier improvements in communications technology.

* * *

Eighteenth-century improvements in roads were largely financed and run by turnpike trusts and, just like dot com companies, the trusts were

[16] Kaukiainen, Y. (2001) Shrinking the world: Improvements in the speed of information transmission c. 1820–1870. *European Review of Economic History*, Vol. 5, pp.1–28.

[17] Collins, F. (2001) *The first information superhighway*. BBC History Magazine, September 2001, p.58.

[18] *Quantum Information*, Visions Paper 2, p.1. Institute of Physics, London (no date, c.2000).

typically small, consisting of perhaps a dozen people and having responsibility for just 20 or 30 miles of road in total.[19]

The growing trade in transporting people and goods even spawned a new breed of criminal, just as e-commerce has spawned a myriad of computer hackers. Hackers are often portrayed as 'a select group' of 'elite' criminals,[20] and eighteenth-century highwaymen were seen in much the same way. The best known is Dick Turpin who, despite having killed a man and having been hanged in 1739 as a notorious criminal, was later the subject of much 'romaticization'.[21] Even in his own day, he was seen as a higher class of criminal – an anonymous pamphleteer, writing in 1736 about the idea that drinking cheap spirits led to criminality among the working classes, said that 'as to highwaymen, experience shows they are of another sort, more properly to be ranked amongst the wine-drinkers.'[22]

The railway boom also resembled its modern internet counterpart. Once again, there was soon a plethora of small companies hoping to operate profitable services. In 1836–7, forty-four companies, concerned with 1,498 miles of track, were sanctioned in Britain. But this rapid growth was to be eclipsed by the 'railway mania' of the next decade. In the period from 1845 to 1847, 576 companies, and a further 8,731 miles of track were approved by Parliament.[23]

* * *

But although the turnpike trusts, railways and radio technology made a significant contribution to improvements in the flow of people, goods and information around Britain and the world, they were by no means always good personal investments. The managers of the Totnes Turnpike Trust in Devon habitually spent more money each year than they raised, and were frequently borrowing large sums. Between May 1780 and December 1781, they overspent by 13 per cent of their turnover, and between 1822 and 1845, expenditure exceeded income in 12 years out of the 14 for which records survive. When the Trust was eventually wound up, £28,300 was owed to the shareholders, but there was only £7,325 in the accounts.[24]

[19] Rule, J. (1992) *The Vital Century: England's Developing Economy 1714–1815*, p.218. Longman, London.

[20] *The Times*, 1 August 2001.

[21] Gardiner, J. (ed.) *The History Today Who's Who in British History*, p.792. Collins & Brown, London.

[22] *A Proper Reply to a Scandalous Libel, Intituled The Trial of the Spirits, In a Letter to the Author*, p.12. Second Edition, with additions, London, 1736.

[23] Briggs, A. (1994) *A Social History of England, New Edition: From the Ice Age to the Channel Tunnel*, p.231. Weidenfeld & Nicolson, London.

[24] Lowe, M.C. (1987) *Turnpikes and Tollgates: A Study of the Totnes and Bridgetown Pomeroy Turnpike Trust 1759–1881*, pp.6, 13, 32, 43 and 50.

At the other end of the country, Captain Robert Barclay took over the coach service from Edinburgh to Aberdeen. He cut two and half hours off the journey time, and his service became 'the model for speed, punctuality, efficiency, smartness, courtesy and service'. Barclay's running of the coaches was so highly praised that in July 1835, a testimonial dinner was held in his honour. But the economic return did not match the effort, and Barclay's Edinburgh to Aberdeen coach service was never a commercial success.[25]

The railway boom was a little different from earlier information revolutions, and there is no doubt that investing in the railways could be highly profitable. Thomas Brassey, one of the greatest railway contractors anywhere in the world, left a fortune of more than £3 million when he died in 1870.[26]

However, the capacity for economic failure among railway companies was enormous, and began to be demonstrated at a very early stage by the first railway company in London. Approved by Parliament in 1833, it was intended to have its terminus at Greenwich. The line was still not finished by 1837, by which time the London and Greenwich Railway Company had run out of money. The Managing Director was fired, as the shareholders wondered what had happened to the fortunes they had been promised when they invested. S.A. Sekon, writing with the benefit of hindsight one hundred years later, expressed the situation by saying 'the capital was exhausted, whilst as all but the credulous optimists expected, the promised dividend failed to materialize'.[27]

In the nineteenth century, the average time from the formation of a railway company until its collapse or takeover was just a few years.[28] By 1912, there was three times as much railway track in the UK as there had been in 1850, but among the operating companies, many were uneconomical and closed rapidly.[29]

Investment in the early American railroads followed much the same pattern, though there were even greater opportunities in North America than in England. Between 1847 and 1890, the length of railway track in the UK expanded by a factor of five, but in the USA, it grew thirty-fold, to over 166,000 miles.[30] Nevertheless, many of the early railroad companies

[25] Radford, P. (2001) *The Celebrated Captain Barclay: Sport, Money and Fame in Regency Britain*, pp.273, 278.

[26] Briggs, A. (1994) *A Social History of England, New Edition: From the Ice Age to the Channel Tunnel*, p.232. Weidenfeld & Nicolson, London.

[27] Sekon, G.A. (1938) *Locomotion in Victorian London*, p.130. Oxford University Press, Oxford.

[28] Hawkings, D.T. (1995) *Railway Ancestors: A Guide to the Staff Records of the Railway Companies of England and Wales 1822–1947*, Appendix. Alan Sutton, Stroud, Gloucestershire.

[29] Royle, E. (1987) *Modern Britain: A Social History 1750–1985*, p.11. Edward Arnold, London.

[30] *The Principal American Railroads: Their History Position and Prospects*, p.20. G. Gregory and Company, London (1892).

proved financially disastrous for the speculators who invested in them. Just like investors in many modern internet companies, most of these people were of 'the middling class of the community' according to Henry Williams, one of the directors of the Boston and Worcester Railroad which was opened in 1837.[31]

In 1834, the USA had fostered such a strong economy that the federal government had paid off the entire national debt, something no other modern nation had ever done. But railroads were sucking up cash and, by 1838, the individual States had huge public debts, including US$43 million attributable to the railroads (and a further US$7 million to the declining organizations that ran the turnpikes). The situation was so desperate that the federal government either could not or would not underwrite the debts, so that the whole nation was given a bad credit rating. Even the government in Washington could not float a loan in Europe, one of the traditional ways of raising money to finance the impressive expansion of the United States economy.[32]

Of course, some railroad investments did prove profitable. People who had loaned money to some companies in the form of bonds eventually saw their stakes converted into shares in profitable companies, even in Ohio, Illinois and Indiana, where some of the major banks had earlier been bankrupted by poor investments in railroads. Thousands more, however, suffered substantial losses.

Although the cost of building railroads had reached US$300 million by 1850, the average length of track run by individual companies was just 36 miles even in the relatively easy terrain of New England, and investors found their assets to be 'unremunerative'. Around the country, companies like the Harlem Railroad and the Hudson Rover Railroad were being described as 'sorry business ventures' and, in the 1850s, the prices of railway stocks and shares quoted in *Hunt's Merchants' Magazine* were more often than not lower than the prices that shareholders had paid for them.[33] In 1853, the US Treasury made a conservative estimate that there were some US$480 million worth of railroad stocks and bonds outstanding, of which over US$50 million was foreign-owned.[34]

[31] Daniels, W.M. (1932) *American Railroads: Four Phases of their History*, p.9. Princeton University Press, Princeton, NJ.

[32] Adler, D.R. (1970) *British Investment in American Railways 1834–1989* Edited by Muriel E. Hidy, pp.7, 10 and 15. University Press of Virginia, Charlottesville, VA.

[33] Daniels, W.M. (1932) *American Railroads: Four Phases of their History*, pp.4, 10, 19 and 24. Princeton University Press, Princeton, NJ.

[34] Adler, D.R. (1970) *British Investment in American Railways 1834–1989* Edited by Muriel E. Hidy, p.23. University Press of Virginia, Charlottesville, VA.

The warning signs did not deter investors even when financial difficulties reached the USA in 1857, and 'spectacular bankruptcies' were witnessed among the railroads and associated companies.[35] According to the contemporary commentator, J. Edward Meeker, both the stock and bond markets were for the next thirty years 'pre-eminently markets for railroad securities'. Many shareholders suffered as a result and closures, takeovers and failures 'carried into bankruptcy and receiverships a substantial part of the country's mileage'.[36]

The radio revolution followed the same pattern. A large number of companies were formed in the 1920s to manufacture radios and some of them, such as Marconi, Ekco and Pye, proved to be sound investments. But many of the earliest pioneer companies began to drop out of the race once the giants started to corner huge sections of the market. Companies such as BTH, Burndept and Efescaphone, which had been trading from the beginning of radio, had already ceased manufacture by the late 1920s.[37]

* * *

John Law's creation of a popular French bank in the eighteenth century is a superb historical example of how economic failure can easily follow an attempt to generate wealth by using information technology. The Banque Générale issued paper money as a rapid way of transferring capital, convenient for both the rich and middle classes. At the Banque, John Law asserted in 1716, you could transfer money from Paris to the provinces, cash cheques and exchange foreign currency for little or no charge.[38] By using printing technology and modern transportation, he claimed to be revolutionising financial services, just as the HSBC bank told its customers in 2001 that 'Internet banking makes it easy' and that 'it's so simple . . . to be in control of your money'.[39]

Just as there is no doubt that the internet is indeed changing our banking habits, so there can be no doubt that Law was right to believe that paper money was a vastly more efficient way of moving capital than metal coinage, given eighteenth-century technology. The Scottish explorer,

[35] Adler, D.R. (1970) *British Investment in American Railways 1834–1989* Edited by Muriel E. Hidy, p.22. University Press of Virginia, Charlottesville, VA.

[36] Quoted in Daniels, W.M. (1932) *American Railroads: Four Phases of their History*, pp.16 and 25. Princeton University Press, Princeton, NJ.

[37] Posen, P. (1995) The development of the domestic radio receiver, pp.128–132 in *International Conference on 100 Years of Radio*, Conference Publication Number 41. Institution of Electrical Engineers, London.

[38] Gleeson, J. (1999) *The Moneymaker*, p.104. Bantam Books, London.

[39] *Internet Banking: Get On*. Leaflet issued by HSBC Bank plc and available at UK branches in 2001.

James Bruce, spent several years in Ethiopia between 1768 and 1772, and sent many letters and papers home. The only one that ever arrived was a bill for £300 that was cashed for him by a Greek trader in northern Ethiopia when he ran out of money. In consequence, Bruce recommended that all travellers in remote lands should 'tack bills of exchange to their letters of greatest consequence, as a sure method of preventing their miscarriage'.[40]

Whatever the strength of his argument, John Law was, in fact, an addictive gambler, who had taken a huge risk in setting up the Banque Générale. The Banque had only functioned effectively at all because of state support – the Duke of Orleans, as Regent of France, had publicly turned up on the doorstep, with his servants carrying boxes full of a million livres' worth of gold and silver coins.[41] Even with this sponsorship, the project was doomed to failure, and in 1720, the bank closed down. Many of the shareholders (who had previously been receiving lavish dividends), and even more depositors, lost everything. Law became 'an object of popular hatred' and fled France.[42]

His ambitious talk, of using new methods to reform banking services had come to nothing, partly because he was arrogant, partly because he was a Scotsman trying to tell the French how to run things, but mostly because not all plausible ideas for using new technology are good ones, and even when they are, it generally takes a great deal of trial and error to make them work. The truth is that history is littered with evidence of a high failure rate among the majority of enterprises that are set up to exploit new technology.

* * *

The question for those of us who are concerned to ensure the future security of our pensions is whether or not the financial fortunes of modern high-technology enterprises will be any different from those of the companies set up to exploit the technologies associated with previous information revolutions.

In fact, there is no reason to suppose that the financial performance of the average internet company will be any different from those of the average railway company, the average turnpike trust or the average radio company. Exactly the same set of rules applies to the internet as applied to each previous information boom. Most turnpike trusts never made a

[40] Bredin, M. (2001) *The Pale Abyssinian: A life of James Bruce, African Explorer and Adventurer*, p.202. Flamingo, London.

[41] Gleeson, J. (1999) *The Moneymaker*, pp.93–94. Bantam Books, London.

[42] Magnusson, M. (ed.) (1990) *Chambers Biographical Dictionary*, p.868. Chambers, Edinburgh.

profit, and neither will most internet companies. Most railway companies were swallowed up by larger concerns and most radio companies folded soon after they started trading; something similar is sure to happen to most modern high-technology enterprises.

Even as nine new technology-based companies entered the *Financial Times* 100 Share Index in the spring of 2000, there were signs that the predicted boom in internet stocks would not really materialize. The Dow Jones index of the American stock markets saw its largest ever daily rise, with 'traditional manufacturing stocks to the fore as investors fled the new-economy wonderstocks'.[43] On the very day the new companies entered the London stockmarket's most important index, it was reported that 'shares in all the new entrants have fallen as speculators and pension funds steered clear, fearing that the high-tech bubble might burst'.[44]

Within six months, the *Financial Times* 100 Share Index was evicting a group of companies that no longer met the criteria for membership. Head of the list was the internet company Freeserve.[45] A year later, Japan's 'ailing technology sector' was 'bracing itself for yet more bad news' as 20,000 jobs were cut by two large electronics companies.[46]

Not surprisingly, the predicted financial revolution of the internet has so far failed to materialize, and a great many investors have failed to get rich.

The 'fall of the internet' was said to have occurred when the publishers of a magazine called *The Industry Standard* ceased publication in August 2001. Described as 'one of the most glittering symbols of the high-tech revolution', it specialized in reporting on news about internet companies. Its demise was said to be linked to 'general disaffection with the Net' and it occurred when 'the list of dot com casualties . . . [had reached] the hundreds'.[47]

* * *

For those of us who are making our pension contributions today, and who want to look forward to decent pensions in the future, there seems to be a puzzle. According to just about everybody, the economy is changing out of all recognition, and good returns on our money will come only by investing in the high-technology companies that will be effective in the information revolution, and form part of the knowledge economy. But the examples of the railways, turnpike companies, radio transmission

[43] *The Independent*, 17 March 2000.
[44] *The Times*, 20 March 2000.
[45] *The Times*, 4 September 2000.
[46] *The Times*, 27 August, 2001.
[47] *The Independent*, 18 August 2001.

companies and others, suggest that while information revolutions do bring overall economic benefits, financial failure is just as likely as success for each individual company, if not more so.

To confirm any doubts we might entertain, the press is constantly reminding us that internet companies are not such good investments as we believed a few years ago. The 'fall from grace of telecoms and dot-com companies' caused *The Times* to express fears about volatile stock markets,[48] while *The Independent* thought that new-technology stocks were 'shunned and devalued' and that it was time to 'dump' them.[49]

The effects of this kind of failure in the high-technology economy were made extremely plain by a survey in August 2001, which prompted the *Daily Mail* to run the headline 'How safe are big company pensions?'[50] The survey by independent actuaries Bacon and Woodrow showed that even as they invested in the new economy, seventeen of Britain's largest 100 companies had pension schemes that were unlikely to be able to afford to make the pension payouts that they had already promised to their employees.[51] The high street retailer Marks and Spencer indicated that it would be making up the shortfall from its own resources, thus reducing the funds available to pay dividends to its shareholders, and probably causing a fall in its share price.[52]

* * *

The truth is that while many high-technology companies are bound to fail, the economy is buoyed up in the long term not merely by those companies that seek to *use* existing new technologies, but by those enterprises where the ethos is to *create* new products and processes, by carrying out research and development. This was true of radio companies and those responsible for eighteenth century roads. Marconi, one of the few radio companies that blossomed after the late 1920s, did so by constant innovation. For example, Guiglelmo Marconi himself was the first person to realize the potential of microwaves for communication, and in 1932 installed the first operational system, between the Vatican and the Pope's summer residence at Castel Gandolfo.[53]

[48] *The Times*, 20 March 2000.

[49] *The Independent*, 24 March 2000.

[50] *Daily Mail*, 15 August 2001, p.1.

[51] See <www.bacon-woodrow.com> for the details of the report, which looked at the pension schemes of the FTSE 100 companies.

[52] *Daily Mail*, 15 August 2001, p.4.

[53] Olver, A.D. (1995) Trends in antenna design over 100 years, pp.83–88 in *International Conference on 100 Years of Radio*, Conference Publication Number 41. Institution of Electrical Engineers, London.

Likewise the great success of turnpike trusts was the legacy of John Loudon McAdam, who decided not just to operate the existing road system, but to engage in engineering research. When he lived in Ayrshire in Scotland, he performed a series of experiments that led to his development of a new road surface. Appointed as surveyor of the Bristol Turnpike Trust in 1816, he remade the roads using crushed stone bound together with gravel, and raised their height to improve drainage.[54] Roads built in this way were known as macadmized roads, and McAdam's legacy flourished with the invention of 'tar macadam' during the second half of the nineteenth century. A company called Tarmac continues to be a successful road builder, with an annual turnover of more than US$4 billion, although its origins do not extend back as far as John McAdam's lifetime.[55]

The drive to research and develop new products and processes is still the force that promotes economic growth in high-technology companies. Chapter 1 described how the modern car companies that grew successfully over the 1990s were those that continued to invest in research and development, even during the financial hardships of the recession of the early part of the decade. The two biggest investors in research, Toyota and Honda, created almost 100,000 new jobs between 1992 and 1999 while Fiat was losing jobs, following its failure to carry out substantial amounts of research.[56]

The success of the NASDAQ stock market, which specializes in technology-based companies, has focused attention on the investment potential of science-based enterprises. The market value of a company, calculated as the total value of all its shares at the current price, is generally much greater than its book value, which is basically the price that might be commanded if all the firm's existing assets were sold off. The difference between the two gives a rough guide to the degree to which investors are confident that profits will be generated by the company. If these profits are actually generated, they are either paid as a dividend to shareholders, or contribute to an increase in the share price.

Analyses of the world's markets in stocks and shares show that market value is much higher for those companies that carry out research and development than for similar companies that do not. The larger investors, like the large pension funds, recognize that research will generate new ideas, which in turn will generate new products and processes that will

[54] Magnusson, M. (ed.) (1990) *Chambers Biographical Dictionary*, p.931. Chambers, Edinburgh.

[55] *Tarmac Review 98: Annual Report 1998.*

[56] See page 11, supra.

keep science-based companies ahead of the competition.[57] M.R. Tubbs followed the fortunes of the world's companies and calculated how much money he would have been able to make by investing in research-based companies compared to the profits that would have been generated by investing an equal amount in a 'tracker' portfolio of shares, similar to those used by many of the companies that handle the relatively small amounts of money available to private investors.

Between 1994 and 1999, the tracker funds did well, because the world's economy was booming. But by the end of the period, the portfolio of research-based companies was worth almost twice as much as a typical tracker fund. At the beginning of 2000, the research-based portfolio had soared in value, and was worth more than five times as much as the tracker funds, but later in the year, the financial markets realized that some high-technology companies were overvalued, and the tracker fund gained some ground against the research-based fund. Nevertheless, Tubbs's experiment showed very clearly how companies that take research seriously really are worth more than those that do not.[58]

The message for all of us is clear. As we move towards less and less state provision of pensions and benefits in our old age, we can be most confident that our own pension funds will grow if the companies which manage our funds choose to invest not merely in companies that seek to use new technologies, but focus to some extent on those companies which, by carrying out extensive research and development, will invent the economic success of future decades.

[57] A summary of the research in this area is given in *The 2000 R&D Scoreboard: Commentary and Analysis*, pp.1–23. Department of Trade and Industry, London (2000).

[58] *Investors' Chronicle*, Vol. 131, No. 1672, pp.22–24 (2000).

3 Politics, the law and selling us dog food by using statistics

The use and abuse of statistics in advertising, political announcements and legal cases highlight the need for us all to be wary when others try to use science and mathematics to convince us of their arguments

MOST of us probably do not buy many things just because we see a good advertisement. We might laugh at the adverts if they are funny, or be impressed if they have striking images, but we are rarely so convinced by the advertisers' arguments that we rush out to buy whatever it is they are trying to sell. Nevertheless, companies keep spending large amounts of money on advertising, and it is difficult to believe that they do not have some kind of evidence that what they are doing is effective, even if it is only making sure that we continue to recognize their goods and services.

Between 1999 and 2000, the British television network, ITV, brought in advertising revenue of £354.4 million.[1] This represented an annual increase of £100 million and Leslie Hill, the Chief Executive of the network, attributed as much as £70 million of this increase to the fact that fees had been raised for companies wanting to advertise in the peak viewing hours between 7pm and 10.30pm. He said that advertisers were paying 'quite substantially' more to advertise because the network had increased its share of the available audience.[2]

In part, the higher ratings were generated by the company's decision to move its main news bulletin from 10pm to 11pm, opening up the prime-time schedules for more blockbuster films, dramas and comedy programmes which attract large audiences, and hence large advertising contracts. Pushing the news back to a less popular slot in the schedules angered many politicians. Because politicians often appear on the news, they like to think that everyone should watch it. Led by the outspoken member of Parliament, Gerald Kaufman, they criticized the television company

[1] *Carlton Communications plc: Annual Report 2000.*

[2] *Minutes of Evidence of the Culture, Media and Sport Committee of the House of Commons*, 2 March 2000, question 2.

for 'throwing away' two million viewers 'for the sake of a . . . boost in advertising cash'.[3]

The same company was in trouble again in October 2001 when it was forced to move its flagship sports programme, *The Premiership*, because only five million people watched it. One Saturday evening, just 3.1 million people watched the show while 6.8 million watched a quiz show on BBC1, the main rival channel.[4] David Liddiment, the head of ITV's broadcasting channels said that pressure on advertising revenues meant that the company 'simply cannot sustain the current position'.[5]

The television bosses are not stupid to worry about advertising income. There is a great deal of money being spent. Over US$90 billion is spent on advertising in Europe each year.[6] In the USA, advertisers spend about US$100 billion each year,[7] and by 2012, they will probably be spending about US$33 billion annually on television advertising alone.[8]

* * *

In their attempts to sell us things, advertisers use science all the time. There was even a television advertisement for L'Oreal shampoo in which the famous American actress, Jennifer Aniston, told us to pay attention because 'here comes the science bit'. Miss Aniston is popular because she is extremely beautiful and because she appears in *Friends*, a situation comedy which has achieved startling success. In fact, she is so well known that at one time, women would ask their hair stylist to give them 'a Jennifer' if they wanted to copy her coiffure. She is not known for her scientific training, and her words in this advertisement were intended merely to convey the idea that some boffins (who were neither as physically attractive nor as famous as Jennifer) had formulated a product that would be gentle on our hair, and get it clean, and strengthen it, all at the same time. It was not enough to give the impression that by washing with L'Oreal, we too could be gorgeous and famous; the company thought that we would appreciate knowing that the shampoo was backed by science.

Graphs of one form or another are particularly popular scientific tools in advertising. One particularly good example of the pointless use of

[3] BBC News Online, 2 March 2000.

[4] Figures from the Broadcasters Audience Research Board, <www.barb.co.uk>

[5] BBC News Online, 22 October 2001.

[6] Koranteng, J. (1999) *The Advertising Industry in Europe: Evolution Towards the 21st Century*, p.13. FT Media, London.

[7] *World Advertising Trends 1998*, p.236. NTC Publications, Henley-on-Thames (1998).

[8] Waterson, J. (2000) *Long Term Advertising Expenditure Forecast*, p.40. NTC Publications Ltd, Henley-on-Thames. The figure assumes that the per capita rate of spending in the USA will be the same as that in the UK.

graphs came in an advertisement for a kind of dog food known as 'Iams', which claimed to be a 'balanced diet that works with the changing immune system of your dog – day by day, in every stage of his life'.[9] Accompanying the claims was a graph purporting to show how an elderly dog's immune system was strengthened by eating 'Iams Senior'. The axis was labelled 'Strength of Immune System' and one bar represented the strength of the system in an eight-year-old dog that had not received the benefit of the special diet, while a second bar represented the equivalent strength of the system of an eight-year-old dog that had been fed on the formula diet for eight weeks. The second bar was more than half as long again as the first bar, from which we were presumably supposed to infer that 'Iams Senior' strengthens the immune system by at least 50%.

But the graph is actually meaningless. The axis contains no units of measurement. We are given no idea at all of what has been measured. Was the strength of the immune system measured using blood tests to see how many antibodies the dogs were capable of producing? Was an experiment conducted in which dogs were deliberately infected with infectious diseases and those that were not fed 'Iams' took 50% longer to recover? Or did 50% fewer of the dogs on the special diet die from infectious diseases?

Any of these might be interpreted as 50% stronger immune systems, but none of them seems particularly likely to have happened. In fact, there is absolutely no information at all in the advertisement that gives any indication of what the claims about immunity are supposed to mean. There is not even any scale on the axis, so it is not even clear whether or not the lengths of the bars are supposed to be proportional to one another, or whether, in fact, feeding your dog with Iams makes only a marginal difference to the elusive measure of immune strength.

Presumably, the pet food that is being advertised is genuinely good for dogs, and has been shown to be so. Otherwise, the advertising standards authorities would have been able to take action to stop the company from running the advertisements, and might possibly have fined them. But the advertisement itself tells us nothing useful about the health-giving benefits of 'Iams' dogfood.

* * *

Mathematical information is among the most commonly used scientific weapons in the armoury of the advertisers. In using statistics, marketing people often seem either to abuse the numbers in an effort to mislead us, or perhaps they actually misunderstand the statistics themselves. The

[9] *Now*, 24 October 2001, p.23.

American mathematician A.K. Dewdney points out that because prices are numbers, that they can be 'minimized, underestimated, or detached from context'.[10]

The simplest example is the common practice of telling us something like 'This car is just $150 per month!' without bothering to tell us how many months it will take to finish paying. Among the worst of these abuses is probably the custom of debt-management companies which offer to compound all our debts into a single debt, and then reduce our monthly outgoings. Their advertising never draws attention to the extended period over which we will have to pay off the debt.

Because we know that these debt-management companies are commercial concerns, the whole purpose of which is to make money, a moment's cool thought will convince us that we will end up paying out more money by using their services than by not doing so. However, since we are likely to be desperate if we have large debts, we might find that a moment's cool thought is extremely difficult, and we might be particularly vulnerable to this kind of advertising.

A leaflet distributed in 2001 by a company called MBNA asks the reader to 'Stop! Paying out so much on your existing borrowings' and 'see how much you could save every month'. It gives an example of someone with existing debts of £5,000 made up of a bank loan, an overdraft, a credit card bill and an outstanding store card balance. If this person takes out a loan from MBNA for £7,500, he or she will have £2,500 in cash to spend immediately, and new monthly repayments of just £160.80. The advertisement fails to point out (except in very small print at the bottom) that these payments must be made over a period of five years, and that the total money repaid will be £9,648 – almost twice as much as the original £5,000 debt.[11]

This example appears, at least on the face of it, to be a deliberate attempt to use numbers in a way that, although strictly accurate, has considerable potential for confusion. Other adverts appear to use numbers in ways that do not seem to be outrightly misleading, but which nevertheless have the effect either of confusing us or making it difficult to envisage the full range of possibilities with which their statistics might be consistent.

One such advertisement appeared in English magazines in about 1999, attempting to convince us that chocolates from the Sainsbury's supermar-

[10] Dewdney, A.K. (1993) *200% of Nothing*, p.43. John Wiley and Sons, New York.

[11] The small print in the leaflet, which was distributed with a mailing from a book club, points out that the example 'assumes that no charges are incurred' without giving any indication of how likely it is that they will be incurred. Furthermore, the repayment figure assumes that the borrower will not take out 'Payment Protection Cover', which would increase the total money repaid to more than £11,000.

ket chain tasted better than those sold in other supermarkets. It began: 'Over the last year, Sainsbury's carried out 90,000 taste tests. In one test, nearly 70% of customers preferred the taste, look and variety of our Belgian Chocolates to another quality retailer's.' This piece of mathematical marketing is so ambiguous that it is useless in helping us to decide whether we want to buy these chocolates.

One possible reading is that the company carried out 90,000 taste tests on chocolates and that, on one of those occasions, customers preferred Sainsbury's chocolates so that, in 89,999 tests, the competitor's chocolates were deemed nicer. This cannot be correct because it would imply that Sainsbury's chocolates were completely disgusting to taste, hideous to look at, and utterly unsatisfying in their variety. Even if this were true (which it obviously is not), the company would hardly advertise the fact.

We have to accept, then, that the 90,000 taste tests covered an assortment of different foods sold in Sainsbury's supermarkets. But without any further information, we have no idea whether there was one test on each of 90,000 different products, or 45,000 tests on each of two products, or 10,000 tests on each of nine products, or three tests on each of 30,000 products, or any one of thousands of other possibilities. We know that modern supermarkets sell many thousands of different product lines, so the customer is probably safe in assuming that the company tested a large number of products, and carried out relatively few tests on each product.

We still cannot guess, however, exactly how many taste tests were performed on Belgian chocolates, and it is perfectly plausible that there was a single test or that there were five or ten tests. This information is important, because we know that there was just one test in which nearly 70% of tasters preferred the chocolates from Sainsbury's shops. If this result occurred in the only test that was carried out on chocolates, it may be far more significant than if it occurred once out of ten tests.

Statisticians concern themselves with 'sample sizes' a great deal, but everyone can see that the larger the sample that we observe, the more likely we are to obtain a particular result – we stand more chance of winning the Lottery if we buy ten tickets than if we buy just one. The more tests the supermarket performed, the less astonished we should be that one of the tests produced a result that gave the impression that Sainsbury's chocolates were superior to those of other stores.

As it happens, it still would not help us a great deal to know how many chocolate tests were carried out to produce one positive result, because at no point are we told how many people were involved in each test. We know that 'nearly 70%' of customers preferred the Sainsbury chocolates in the particular test that we are invited to notice, but we have no idea whether this means 13 people out of 19 (which is 68%) or 69 people out of 100 (69%). All in all then, the Sainsbury's advertisement is utterly meaningless. It seems astonishing that somebody in the marketing department of the company thought that somehow it would sell us more chocolates, and it is difficult to believe that anyone actually bought any of Sainsbury's confectionery as a result of the advert.

This kind of statistical smoke-and-mirrors is unpleasant, confusing and pointless, but as individuals, we can avoid it if we choose. However, there are two situations where statistical ambiguity and abuse, of the kind that occurs in advertising, are important in terms of public policy – politics and the law. In a democracy, political parties are essentially advertising for our support and votes, and in a legal case, each side is advertising for the support of the judge and jurors. Both politicians and lawyers often try to blind us with numbers.

* * *

When the Labour Party came into power in the UK in 1997, the government moved away from the system of annual haggling over the budgets of each ministry, and towards a system of three-year public spending announcements. Ministers made a strong case for the benefits of explicit, known, forward-looking plans, but the system opened up the possibility of huge misunderstanding in the way in which the numbers were presented.

The problem arose with the announcement of the first 'Comprehensive Spending Review' in the summer of 1998. The budget of the National Health Service was set to rise from £36 billion in 1998 to £45 billion in 2001, but there were so many different ways of presenting these figures that it became almost impossible to understand what was really happening.

Gordon Brown was the Chancellor of the Exchequer who announced the figures, and who wanted them to look impressive. He tried to express the increasing budget simply, by saying, 'I am announcing an increase in health service funding of £21 billion'.[12] Some people saw this as 'triple accounting' because the money was to spread over three years. In other words, there was to be an average of £7 billion of new money each year.[13]

The official publications that detailed the figures presented the announcement in percentage terms. They reported that there would be 'average real increases of 4.7%, in real terms' in the Health Service

[12] *Hansard* [House of Commons], 14 July 1998, column 194.
[13] *The Times*, 22 March 2000.

budget for each of the coming three years.[14] But this might be construed as misleading because the actual rises were 8.4%, 7.2% and 6.5%, all of which are higher than the quoted average. The difference is partly accounted for because the average figure is the 'real terms' growth. This calculation assumes that inflation will increase prices by a certain amount every year, so that the same quantity of cash will not buy as much in three years' time as it will today.

None of these methods mentioned that the annual budget of the National Health Service would be £8.7 billion higher at the end of the budgetary period than it was at the beginning. This is what was meant by Francis Maude, the opposition spokesperson, when he accused Gordon Brown of 'a series of accounting fiddles' and challenged the figure of £21 billion, saying it was 'actually £9 billion'.[15] The morass of figures was utterly confusing, even though each set was, strictly speaking, accurate and defensible, according to one interpretation of the truth.

In fact, none of the reports and comments gave the figure that might actually have been most useful. Since few people can think about millions or billions of pounds without becoming confused, it might have been more meaningful to tell us the overall proportional increase in the budget of the health service. Taking into account the predicted effects of inflation, there was to be 16% more buying power in the budget at the end of the three-year period than there was at the beginning.[16]

Exactly the same problem occurred in March 2001. While the Chancellor of the Exchequer, Gordon Brown, announced that £1 billion was being added to the budget of the National Health Service over three years, his political opponent, Michael Portillo, said, 'there isn't a billion for health . . . [Mr Brown] . . . is trying to con people.'[17] The *Daily Telegraph*, taking its lead from Mr Portillo, calculated that, since the £1 billion was spread over three years, funding was actually rising by an average of £333 million per year.

The problem for the rest of us is that we do not know whose version of events is the best representation of reality. They are all telling the truth, in the strictest sense but, in most cases, we need a detailed set of caveats and definitions to understand each different version of the truth. In politics, this matters because it is difficult and unrealistic for individual citizens to

[14] *Comprehensive Spending Review: New Public Spending Plans 1999–2002. Modern Public Services for Britain: Investing in Reform*, p.46. The Stationery Office, London (1998).

[15] *Daily Telegraph*, 15 July 1998.

[16] This figure uses the standard Treasury assumption of 2.5% inflation per year.

[17] *Daily Telegraph*, 9 March 2001.

wade through the technical detail of each set of announcements in order to make a sensible judgement about how statistics are being used.

However, the effects of these difficulties are ameliorated in a free democracy because pressure groups and journalists make it their business to get to the bottom of apparent disparities, and to present the information in ways that help us to understand the underlying truth. Moreover, we can make further judgements, based on our own experience. In public services, it is not the amount of money that really matters, it is the standard of service that we receive. If we experience overstretched, understaffed hospitals, we know that policies are failing, and it is irrelevant whether there is £21 billion or £9 billion involved. When statistics are brought into legal cases, however, the potential for harm may be much greater.

* * *

In the legal system, the effects that temper the misuse of statistics are less powerful than they are in politics. Although pressure groups and journalists may take up the cause of an individual who appears to have been wrongly convicted on the basis of a misleading interpretation of statistics, they cannot monitor every case, and those who are convicted are likely to suffer in the short or medium term even if they are eventually cleared. And if a dangerous criminal is acquitted by mistake, there is not much anyone can do about it unless the crook re-offends.

Furthermore, judges, and especially jurors, are unlikely to be able to fall back on personal experience in the way the electors can when assessing political statistics. Most of us do not encounter murder, rape and other serious crime, and most of us, including most scientists, understand little about the details of genetics, so that if we are on the jury in a serious criminal trial, we have no basic reference points from which to gauge the merits of the conflicting interpretations of the scientific evidence relating to DNA. It must have been horrendously difficult for the jury in the trial of O.J. Simpson, who had to contend with bizarre facts, media hype, unparalleled public interest and a weak judge, as well as inferences drawn from blood samples.

Legal cases increasingly involve scientific evidence. As science develops new techniques, both criminals and the law enforcement agencies develop new ways of working. When fingerprint identification became possible in the late nineteenth century, it opened a new and powerful method for the police to gather evidence about criminals, but it also offered new ways for criminals to mislead the police. By planting objects, it became possible to implicate innocent people who had never even been present at a crime scene, but who happened to have blamelessly handled the object.

In cases involving complex scientific evidence, such as identifications based on DNA, it is important for the jury to understand the statistical concept of 'conditional probability'.[18] Conditional probabilities can be confusing, but a simple example would be to correlate language and nationality. There is a very high probability (say 95%) that someone who is a French citizen can speak French. But there is a much lower probability (about 20%) that someone who can speak French is a French citizen.[19] He or she may come from another Francophone country, such as Algeria, or may be a native of an entirely different country, who has chosen to learn French.

This type of distinction is important in cases involving genetic identification, because the jury must remember that the question to be asked is 'What is the probability that the accused is innocent, given that his DNA matches a sample from the crime scene?' and that this is different from 'What is the probability that the accused person's DNA matches a sample from the crime scene, given that he is innocent?'

Expressed in this way, it is difficult even for the mathematically-minded amongst us to grasp just how different these questions are, but a fictitious example makes the distinction much clearer. A DNA sample is taken from the scene of a murder which occurred in a city of two million people. Exhaustive tests show that it matches two of the city's inhabitants. This represents a remote, but realistic, one-in-a-million chance of a match.

Ignoring the possibility of the criminal being from out-of-town, we know that one of the two people with the DNA match must be the murderer, while the other is completely innocent and just happens to have a matching genetic profile, according to the tests that the police are able to perform. If one of the two people is accused in court, the prosecution will make much of the fact that only one person in a million has this genetic make-up. They will seek to convince the jury that guilt is strongly indicated by the fact that the accused has genes matching those in the sample from the scene of the crime, and that there is only a one-in-a-million chance of such a match.

The prisoner's defence must stress that the prosecutor is asking the wrong question. His lawyers will argue that the real dilemma for the jury

[18] Although I have changed the details and the precise numbers, the following paragraphs rely heavily on the essay 'DNA Fingers Murderer' in Paulos, J.A. (1996) *A Mathematician Reads the Newspaper*, pp.72–73. Penguin, London.

[19] According to Crystal, D. (1997) *The Cambridge Encyclopedia Third Edition*, pp.423 and 915, Cambridge University Press, Cambridge, approximately 270 million people worldwide speak French as a first or second language, and the population of France is approximately 58 million.

is to decide whether the person is guilty, given that his DNA matches the sample found at the murder scene. We know that two people have DNA matches, so that the probability that either one of them is guilty is one-in-two, or 50%. This represents very much more than a 'reasonable doubt' and unless the prosecution can find some other evidence to link the prisoner with the murder, the DNA evidence is useless – it cannot tell us which is the guilty party out of the two people whose genes match the sample from the crime scene.

Since most of us are not statisticians, including most lawyers and judges, and since lawyers will use every trick they can imagine to win over the jury, it is hardly surprising if individual jurors are not sure what to make of the conflicting mathematical interpretations of this kind of evidence. Legal cases that turn on our understanding of mathematics are not uncommon. There have even been trials involving what the *Financial Times* has called 'a mish-mash of speculative theories and poor statistics', and in some instances the potential for harm is enormous.[20]

In November 1999, a British lawyer, Sally Clark, was convicted of murder, after two of her infant children had died suddenly. When the first child died, doctors had certified the cause of death to be Sudden Infant Death Syndrome, more commonly known as Cot Death Syndrome. But after the second child also died with no apparent cause, the authorities became suspicious, and a police investigation led to Mrs Clark's arrest, trial and conviction.

The jury had been told that the chances of two cot deaths occurring in one family were 1 in 73 million. This appeared to be based on the knowledge that, in the UK, about one child in 8,500 dies without any explanation during their first year. On the assumption that such tragic occurrences are randomly distributed among the population, and that every child is equally likely to be unlucky enough to die inexplicably and suddenly, this figure can be squared, to give the probability of two such events in one family. The square of 8,500 is 73 million, so the probability of two cot deaths occurring in one family was estimated as 1 in 73 million.

The jury convicted Sally Clark. It would not be surprising if the jurors had considered that an event that could legitimately happen to only one family in 73 million was extremely suspicious, and that something untoward must have occurred. Mrs Clark was sent to prison for life. Two years later, the case was reopened, when new research showed that Cot Death Syndrome had a genetic basis, and that, therefore, deaths were unlikely to

[20] *Financial Times*, 18 August 1998.

be spread randomly throughout the population.[21] In fact, they are more likely in families that have already suffered one infant death, and where either the mother or the father might be carrying the relevant genes. This apparently 'new' evidence seemed to cast doubt on the astonishingly small probability that the jury had been led to believe was an accurate estimate of the likelihood of two infant deaths occurring in a single family.

But the statistical problems were not only foreseeable, they were, in fact, foreseen. Three days after Sally Clark had been convicted of killing her two children, a man named Allan Reese had written to *The Guardian* newspaper explaining how it was wrong to treat instances of Sudden Infant Death as independent events. Although he did not indicate why this was so, it is obvious if we accept that *something* is causing the deaths. Just because we do not know what causes cot deaths, it does not follow that there are no causes. The evidence presented in 2001 showed that some babies may be genetically susceptible, and previous evidence had suggested that it might be important whether a baby sleeps on its back or its stomach. It is easy to think of other possible contributing features of the environment, such as some particular kind of cotton sheet, or atmospheric toxin such as cigarette smoke. There is no evidence that these particular things are important, but they illustrate the kind of possibilities that might be worthy of investigation.

Whatever the actual factors, *something* causes cot deaths, and it would be astonishingly unlikely if the important things were distributed entirely randomly. If cigarette smoke or cotton sheets happened to be important factors, two babies in a family of smoking cotton-lovers will both be at greater risk than two babies born into a family of non-smokers who use nylon sheets. But this lack of randomness in the factors causing sudden infant deaths was not Allan Reese's main point. The jury had examined the probability that a family chosen at random would suffer two Cot Deaths, which is quite different from 'the probability that two deaths, once observed, were of that type'.[22]

Because this is not the sort of issue that we normally think about, the distinction between the two ways of looking at the problem is not easy to grasp, but it is a very real one. When the second child died, Mrs Clark's family was not a random family, it was one in which one child had already died, apparently of Sudden Infant Death Syndrome.

The relevant probability seems never to have been put before the court, and Mrs Clark was either innocent or was convicted for the wrong

[21] *The Guardian*, 15 July 2001.
[22] *The Guardian*, 12 November 1999.

reasons (at least in part). In fact, when the case came before a panel of Appeal Court Judges, they felt that mistakes in the way statistics were presented at the original trial had not prejudiced the jury. 'If there had been no error in relation to statistics at the trial, we are satisfied that the jury would still have convicted', they said.[23] Their belief was presumably based on the fact that there was other evidence, besides the sheer improbability of the two deaths, that had convinced the jury of Mrs Clark's guilt.

Another example of dubious statistics determining the outcome of a legal case came in 1998, when the Dow Corning Corporation paid US$3.2 billion in an out-of-court settlement to 170,000 women who claimed to have suffered ill effects after receiving breast implants manufactured by the company. The company did not believe that it was responsible, but its insurance company was said to have accepted that given 'the way the US court system is structured . . . the bill for defending each of the 170,000 cases individually would have come to more than the . . . settlement'.[24]

The company knew that juries would listen to two expert witnesses, giving opposing evidence, and that they were likely to sympathize with apparently suffering women rather than a large corporation. But in the circumstances, the expert witnesses would not have been impartial. At least one expert had already made more than US$1 million from testifying in previous cases relating to the breast implants. It was, apparently, largely irrelevant that 'with each successive report on breast implants, the weight of the evidence (never strong) was collapsing'.[25] The courts were never given the chance to determine the conditional probability: what is the chance that breast implants cause health problems, given what we know about the medical histories of women who have had implants.

In many legal systems, the court is allowed to have an impartial adviser to help the judge deal with difficult matters of law. The UK's courts retain many Latin historical phrases, and they refer to such an expert as an *amicus curiae*, literally translated as 'friend of the court'. This person's job is essentially to call the attention of the court to legal precedents or points of law, which might be important to a particular case. In general, an *amicus* is appointed by the court in cases where there is likely to be a need for some specialist or in-depth legal knowledge, over and above that of the trial lawyers or the judge.

But there is no facility for allowing an expert scientist or statistician, other than as a partisan witness for one side or the other. Courts cannot

[23] *The Guardian*, 3 October 2000.
[24] *Financial Times*, 18 August 1998.
[25] *Financial Times*, 18 August 1998.

appoint an *amicus* to help them with difficult mathematical or scientific matters, even though none of the main participants is likely to be scientifically trained.

The nature of most scientific progress does not lend itself to the idea of a single expert, who can advise the court in an absolute manner, as an *amicus* does on points of law. Any scientific evidence in a court case is either going to be based on fully established scientific principles, in which case the defence will not seek to challenge it, or it will be emerging science, in which case there will, inevitably, be reputable expert scientists who take opposing views about its interpretation. Although a 'statistical *amicus*' may have prevented some mistakes in the case of Sally Clark, individual experts would not be able to give definitive opinions, in the way that judges and other legal experts can give definitive rulings about emerging areas of law.

Interpretation of the law is absolute. Judges may give contradictory rulings, lawyers may give uncertain opinions, but ultimately, a country's supreme court can give an unqualified answer to any question of law. By definition, that court's decision is correct. Science is not like that because its truths are set externally, by nature.[26] A supreme court of researchers may decide that the Earth is flat, but they cannot change the external truth that it is in fact spherical. One group of statisticians may have agreed, broadly, that it was astonishingly unlikely for two babies in the same family to die of Cot Death Syndrome, but they could not change the external reality that something non-random actually caused the deaths. However eminent the advice the court was given when Sally Clark stood trial, the advice was wrong.

When it comes to trusting advertisements, believing politicians, or hearing evidence in court, mathematics and science are only as powerful as our ability to handle them.

STOP PRESS: In January 2003, Sally Clark was freed and her conviction overturned. The Court of Appeal accepted that the original statistics had been wrong. Fresh medical evidence showed that Mrs Clark's second son had probably died from a bacterial infection, proving that cot deaths have real causes, and cannot be treated as random statistical events.

[26] The underlying impartiality of the scientific process is the subject of an entire discipline, known as the 'philosophy of science', of which two of the most important contributions are Kuhn, T.S. (1996) *The Structure of Scientific Revolutions,* Third Edition, University of Chicago Press, Chicago, and Popper, K.R. (1959) *The Logic of Scientific Discovery*, New York. Fascinating though they are, debates about the philosophy of science have no bearing on the fact that scientists are, in general, attempting to uncover truths about nature, and they certainly make no difference to whether or not humans can become ill as a result of eating beef infected with BSE, or whether the sun goes around the Earth or *vice versa*.

4 Vultures, and blackouts in our power supply

Environmental problems show why unpopular research sometimes matters and why international co-operation is essential in science

MOST of us care about environmental research in ways that we do not care about atomic physics or inorganic chemistry. We may think that astronomy is interesting and worthwhile, but we do not always think it is useful. We may think that biochemistry is useful, because it invents new drugs, but most of the time, we just wait for new miracle cures to turn up, and are not especially interested in the details of how they were discovered. By contrast, many of us seem to think that ecological research is, almost uniquely, both useful and interesting.

More than half of us claim we enjoy the wildlife documentaries that present the results of ecological research, while only one in six of us gains any pleasure from arts programmes. Week after week in the autumn of 2001, more than eleven million people – one out of every four British adults – watched a documentary called *The Blue Planet*, which showed amazing footage of life in the oceans.[1] In fact, among the British public, 82% say they are interested in environmental issues, compared to just 60% who claim to be interested in sport, and 54% who have any interest in the politics of their own country.[2] Nor does this represent one of Britain's unusual quirks. The Discovery Channel and *National Geographic* magazine are among the most popular media outlets in the USA.

There are many reasons why we find environmental research more interesting and worthwhile than some other kinds of scientific endeavour. We like animals, so any science about endangered species gets us excited. We enjoy visiting wilderness and countryside, so any research about natural habitats interests us. And of course, many of us feel a duty to protect the environment, and pass it on in good condition to future generations.

[1] Figures from the Broadcasters Audience Research Board, <www.barb.co.uk>.

[2] *Science and the Public: A Review of Science Communication and Public Attitudes to Science in Britain*, pp.70–71. Office of Science and Technology/The Wellcome Trust, London (2000).

But the main reason we care about environmental research is that we have all become scared about impending ecological disasters. Most scientists think that global climate change is happening, and although we cannot accurately predict the effects, different groups of scientists are constantly warning us that floods will become more serious in some areas, droughts will become worse in others, and there will be such damage to populations of animals and plants, and to habitats, that the growing human population will struggle to exist, let alone thrive, in many parts of the world.

We do not really know what will happen over the coming century, but our fears about the environment probably have some foundation. The world's ecosystems provide all sorts of services such as treating our waste, and regulating and controlling the balance of gases in the atmosphere. These services are provided free of charge but, recently, ecologists and economists have tried to estimate their monetary value. The first attempt to calculate the total economic value of ecosystem services was made in 1997 by Robert Costanza of the University of Maryland, together with twelve of his colleagues from around the world. Their best estimate was that ecosystems provide services worth US$33 trillion per year.[3] This number is clearly too large for anyone to appreciate its true meaning, but it is thrown into sharp relief by the observation that it is more than the total Gross National Product of all of the world's countries.

In other words, if we had to pay for the Earth's ecological services, we could not afford them. With this in mind, it is not surprising that we care about environmental science. In principle, then, it ought to be simpler for society to organize its researches into the environment than it can those into physics, engineering, medicine, mathematics or economics. If everybody cares, and everyone is interested, what could stand in the way?

In fact, there are three separate reasons why it is exceptionally difficult to organize, fund and carry out research into the Earth's ecosystems within the framework of modern science.

The first reason is obvious – the world's ecosystem is complex, and it is not simple to find ways of investigating its individual components in a meaningful way. The second cause of difficulties in ecological research is our ambiguous feelings about it. We are often sceptical about ecological investigations because it is not possible to study the environment without interfering with it, and we do not like some of the interventions that are required. The third reason why environmental science can be tricky is that each individual project can seem insignificant, both in terms of

[3] Costanza, R. et al. (1997) The value of the world's ecosystem services and natural capital. *Nature*, Volume 387, pp.253–260.

public policy and in terms of scientific importance; it is only when a large number of smaller projects come together that general lessons can be learned. As a result, the funding agencies are sometimes wary of supporting ecological research.

* * *

Ecosystems are complex because they consist of many thousands of interactions. Even a small garden or park consists of hundreds of plant species, innumerable species of insects and other invertebrates, as well as a good spread of fungi, birds, and mammals. Nobody knows how many bacteria and viruses are present, but there are clearly a great many.

Each species interacts with many others – hawks eat smaller birds, mice, worms and beetles, each of which eats several kinds of plants and insects. The plants rely on symbiotic relationships with fungi, and almost all of the species have viral and bacterial diseases, and other illnesses caused by a variety of pathogens. Everything is affected by the weather and the type of soil.

A larger habitat than a simple garden is unimaginably complex. In the famous Serengeti ecosystem in Tanzania, there are 56 different species of hawks and falcons.[4] Assuming that each of these species of birds of prey competes with only one in ten of the others, there would still be something like three hundred different competitive interactions among these birds. There are presumably also a great many such interactions among the 21 species of antelopes that live in the Serengeti, and an uncountable number among the hundreds of species of plants, or innumerable species of bacteria and micro-organisms.

Humans are a major addition to this complexity. We interact with literally thousands of different species in countless ways, with far-reaching effects. Equally importantly, the interrelationships among all of the other species have an uncountable range of effects on humans. Many of these effects are completely unexpected, and present particular difficulties for scientists engaged in environmental research.

* * *

The vultures of India provide an especially graphic demonstration of the complex interrelationships between human populations and their environments. Vultures are common in many of the world's tropical and subtropical environments, and the Indian subcontinent boasts eight different

[4] Sinclair, A.R.E. and Arcese, P. (1995) *Serengeti II: Dynamics, Management, and Conservation of an Ecosystem*. University of Chicago Press, Chicago and London.

kinds. Of these, four are closely related species known as 'griffon' vultures. It is these that have given cause for concern, especially the white-backed vulture and the long-billed vulture. Between 1992 and 2000, these two species declined by 90% across India and, in many places, they have disappeared altogether. Dr Debbie Pain of the Royal Society for the Protection of Birds believes that the situation is becoming 'progressively worse'.[5] Some estimates suggest that one or both of the two badly affected species could be extinct by 2005.[6]

Nobody knows exactly what has caused the decline of the vultures, but the manner in which the birds are dying gives some indication. Before they die, most of the birds appear to contract an illness, in which they become lethargic and develop a drooping neck. For long periods, the head of an afflicted bird hangs down almost to its feet. After about a month, the weakened vultures fall out of the trees and die. Dr Andrew Cunningham, a veterinary pathologist from the Zoological Society of London, performed post mortem analyses on some of the dead birds and found that all of them had died of enteritis and visceral gout.

The most likely cause of the illness appears to be a virus or bacterium, because the disease is very widespread (it has spread to Nepal and Pakistan), and other possible causes, such as pesticides or other forms of pollution, tend to be much more localized in their effects. 'It's got to be an infectious agent', said Andrew Cunningham.[7]

Many people may take the view that Indian vultures are hardly an issue of great concern, and that their viral illnesses are even less important. Most of us probably think vultures are ugly, and their association with bloody, rotting corpses hardly makes them endearing. Indeed, they have never had a positive image: the earliest known use of the name 'vulture' in English was by Chaucer, who recorded how the Greek sinner Tityus was tortured in hell by a pair of vultures that ripped at his liver for ever.[8] By 1603, according to the *Oxford English Dictionary*, the word vulture had come to mean 'a person of a vile disposition'.

Vultures have not fared any better in recent years. In the Disney film *The Jungle Book*, made in 1967, although the vultures eventually end up helping Mowgli the man-cub, they live in an area where all the trees look dead, and tell Mowgli that 'nobody wants us around'. In Disney's *The Lion King*, made almost 30 years later, vultures and hyenas are the only

[5] BBC News Online, 1 April 2001.

[6] *The Guardian*, 5 October 2000.

[7] BBC News Online, 1 April 2001.

[8] Chaucer, G. *Troilus and Criseyde*, Book 1, line 788 (1374); the story Chaucer relates is from *The Odyssey*, Book 11, lines 576–581.

two species whose members are universally shown as unpleasant and harmful.

Even if we put aside our distaste for these birds, it is easy to argue that there are still many more serious problems for the people of India, Nepal and Pakistan, whose standard of living and quality of healthcare is much worse that those enjoyed by the Americans, Japanese, Australians or Europeans. We might even argue that there are many more interesting and charismatic animals to protect, including those, such as tigers, whose plight can unquestionably be blamed on human interference. The mystery disease does not even affect all Indian vultures. Four of the eight vulture species in the country are completely unaffected, and the populations of only two have suffered seriously.

But it transpires that the kinds of vultures that have been affected were performing an unusually useful service. Like all vultures, they eat the carcasses of dead animals. Fewer vultures means that fewer carcasses are eaten, and those corpses that remain begin to fester and rot in the open air. On top of Bombay's Malabar Hill, the Parsees have a problem. They belong to an ancient religion that believes cremation or burial to be an offence, and so their dead must be left above ground to be picked clean by the vultures. The vultures have disappeared, and Khojeste Mistree of the Bombay Parsee Council said, 'it would be wrong . . . to say we do not have a problem'.[9] Carcasses are rotting above ground.

Elsewhere in India, Hindu villagers are having problems because they rely on the vultures to pick clean the skeletons of their sacred cows. Even if the Parsees and Hindus decided to change their traditional ways, and buried the offending human and cattle carcasses, a major problem would remain because there would still be carcasses of wild animals lying festering on the ground.

With the vultures gone, other scavengers are moving in. Populations of crows are increasing, and the number of semi-wild dogs has shot up.[10] At one pile of carcasses in Rajasthan, a thousand feral dogs congregated to carry out the work that the vultures had formerly performed. These dogs are a major source of rabies, which is on the increase in India. As the dog population increased, Andrew Cunningham observed that 'people are now having to protect themselves with sticks'.[11] Every year, 30,000 people die of rabies in India, and doctors believe that 90% of cases are caused by dog bites. Moreover, the country has a shortage of immunoglobin, a drug that is essential for the treatment of the disease, and the situation has

[9] *The Guardian*, 5 October 2000.

[10] BBC News Online, 5 October 2000.

[11] BBC News Online, 1 April 2001.

become so serious that officials and scientists from the World Health Organization joined with those from many countries to hold a major conference on the subject in the summer of 2000.[12]

As they did so, fears grew that the mysterious vulture disease might spread. Birds in Pakistan and Nepal were increasingly affected, and scientists feared that vultures in Europe and Africa could also succumb. If that happens, the human health problems will be multiplied, and we might find that, astonishingly, our ability to reduce the scale of human disaster could depend on our understanding of the interactions between an unknown virus and a group of ugly birds for which we previously had little or no sympathy.

Moreover, this illness is just one of many emerging infectious diseases that could have significant detrimental effects on biodiversity and human health. Many wildlife species act as reservoirs of pathogens that threaten the wellbeing of domestic animals and humans. In Australia and Papua New Guinea, the hendra virus affects humans, horses and fruit bats, while the bacterium that causes bubonic plague is still spread by rats and other mammals in various countries around the world. There was a major outbreak of plague as recently as the early 1990s.[13]

Even when we can design clever experiments to divulge the secrets of ecology, we must live with the fact that animals, plants and rivers do not respect political boundaries. Our scientific endeavours in this field will only succeed if we organize robust international collaborations.

The complexity of the interrelationships between humans and other species is one reason why veterinary pathologist, Andrew Cunningham, believes that 'the authorities need to treat this with . . . urgency and seriousness . . . You need to work with the international community'.[14] His comments are echoed in every debate about global climate change, which is possibly the greatest environmental challenge that has ever faced humankind.

* * *

None us knows how the climate will change over the next century or what the effects will be and, consequently we do not know how we will deal with those effects. But only a small minority of people now believe that there is nothing to worry about. There is growing evidence that we need

[12] BBC News Online, 6 July 2000.

[13] Daszak, P., Cunningham, A.A. and Hyatt, A.D. (2000) Emerging infectious diseases of wildlife: Threats to biodiversity and human health. *Science*, Volume 287, pp.443–449.

[14] BBC News Online 5 October 2000.

to take some kind of action, either to minimize the likely changes, or to try and find effective ways of living with them.

Whatever actions are taken, it will be necessary for different countries to act together in some kind of coordinated effort. That is why many people were angry with President George W. Bush, when he announced in 2001 that the USA had 'no interest' in implementing the Kyoto Protocol, an international agreement that industrialized countries would, by 2012, substantially reduce their emissions of greenhouse gases.[15] The Protocol, adopted by more than 150 countries around the world, set out ways in which carbon dioxide emissions could be reduced, and in which progress could be monitored and verified.[16]

When he refused to ratify the treaty, Bush was vilified by many people in Europe and by many of his political opponents at home. He became demonized as the 'Toxic Texan', and some Europeans started to call him the 'Climate Killer'. His arguments were shared by the governments of Australia, Canada and Japan, who all agreed that the solutions offered by the Protocol would not work anyway, and would harm their economies, effectively forcing them to hand over money to countries that are worse polluters.

President Bush's aides said that the 'Emperor of Kyoto has no clothes'.[17] Some economic models suggested that if the agreement were fully implemented, as much as US$170 billion would flow into 'the great black hole that is Russia and the Ukraine', and this was considered neither desirable nor politically expedient. There were serious doubts about the 'dubious' scientific basis for the belief that the Protocol's provisions would actually reduce the effects of human-induced climate change.[18]

For all the unfavourable remarks that were made about President Bush, he had a point. It is exceptionally difficult to make realistic estimates of what will happen to the world's climate in the coming century. In truth, even under his more liberal predecessor, Bill Clinton, the United States Senate had voted by 95 votes to zero against the Kyoto agreement.

In fact, some scientists are highly critical of the environmental movement's doom-laden prognostications. Professor Philip Stott of London University's School of Oriental and African Studies pours scorn on 'self-indulgent, post-materialist, eco-chondriac, ciabatta eating' campaigners, believing that the computer models used to predict climate change are less

[15] *The Times*, 19 July 2001.

[16] *Implementing the Kyoto Climate Change Agreement*. Parliamentary Office of Science and Technology, London (2000). [POST Note 147].

[17] *The Guardian*, 27 July 2001.

[18] *The Times*, 19 July 2001.

sophisticated than the computer game *Tomb Raider*.[19] Professor Stott's approach may offend some people, especially those who like ciabatta bread, but his basic point is sound. The challenge of dealing with global warming is as much about inventing ways of coping with the changed climate as it is about futile attempts to turn back the tide and prevent the climatic changes that have already been locked into the Earth's system by the historical production of greenhouse gases.

But our efforts to predict what might happen seem to be improving because of the Intergovernmental Panel on Climate Change, which is an example of effective collaboration between different countries. Consisting of experts from all over the world, it takes a hardheaded view of the science of climate change, and couches its findings in terms of what is 'unlikely', what is 'likely' and what is 'very likely'. In effect, it is open and honest about the level of scientific uncertainty in its conclusions. In a recent report, the Panel was able to conclude that globally, the Earth's temperature has risen by about 0.6 °C since 1861, and that about 10% of the world's snow cover has melted in the last forty years.

Moreover, the international panel of experts is almost certain that the production of 'greenhouse gases' such as carbon dioxide has been an important factor in these global changes. It is, they say, 'likely' that most of the warming up of the Earth's surface over the past fifty years is due to the increase in greenhouse gases. It is 'very likely' that the twenty-first century will be warmer than the twentieth century and 'likely' that there will be an increase in the risk of extreme weather events such as droughts and cyclones.[20]

Science of this nature can only be done in a truly international context. Indeed, Robert May coined a soundbite when he was the UK government's Chief Scientific Adviser, by saying that 'science was globalized long before "globalized" was even a word'.[21]

The great scientists of the past realized the need for their endeavours to take account of a much wider community of researchers. The famous American scientist and inventor, Thomas Edison, visited England to learn about telegraphy from the British Post Office,[22] and Darwin travelled all over the world before reaching his conclusions about evolution by natural selection.[23] Galileo moved from Pisa to Florence, and eventually went to

[19] Kaplinsky, J. (2001) *Better safe than sorry?* Spiked-Online, 20 July 2001. <www.spiked-online.com>.

[20] *Summary for Policymakers: A Report of Working Group I of the Intercontinental Panel on Climate Change*, pp.2, 7, 10 and 15. IPCC, Geneva (2001).

[21] At a Press Conference on 26 July 2000.

[22] Israel, P. (1998) *Edison: A Life of Invention*, pp.83ff. John Wiley & Sons, New York.

[23] White, M. and Gribbin, J. (1995) *Darwin: A Life in Science*, Chapter 3. Simon & Schuster, London.

Padua. His book was 'a sensation throughout Europe', and indeed was only published because it could be smuggled out of Italy to the scientific community in the Netherlands.[24] Einstein was such an international figure that, when he won the Nobel Prize for Physics, it was unclear which country's Ambassador in Sweden should accompany him to the awards ceremony.[25] Science works best in an international context.

* * *

The complexities of dealing with, and paying for, the international dimension of biodiversity and environmental issues have provided the basis for one of the UK government's most successful schemes for the funding of scientific work. Other countries have similar schemes, which have varying success, but few are as highly regarded at the Darwin Initiative for the Survival of Species.

The Darwin Initiative was announced by John Major when he was Prime Minister, following his attendance at the Earth Summit in Rio de Janeiro in June 1992. The scheme seeks to safeguard the world's biodiversity by drawing on national strengths within the UK, to assist those countries that are rich in biodiversity but poor in financial resources.

Projects aim to involve local people, and to ensure that conservation actions fit in with their traditional land management techniques. In the highlands of Ethiopia, a project aspires to understand how and why the local farming practices have allowed the Ethiopian wolf to survive, while it is extinct over the rest of its range. It is the rarest member of the dog family in the world, with only about 500 individuals surviving in a handful of areas in the wild, and almost none in captivity.[26]

In Cameroon and Equatorial Guinea, the Darwin Initiative is trying to understand the implications of the trade in bushmeat, while in Chile it is establishing a programme of monitoring penguins. In Vietnam, it is supporting a community-based conservation project in the Hoang Lien Mountains.[27]

It is also supporting the project to understand the illness that is killing vultures in India, Pakistan and Nepal, which has led to an explosion in the population of rabies-infected dogs that feed on the rotting carcasses the vultures have left. In doing so, it is matching the expertise of Andrew

[24] Porter, R. and Ogilvie. M. (2000) *The Hutchinson Dictionary of Scientific Biography*, Volume I, p.402. Helicon, London.

[25] Fölsing, A. (1998) *Albert Einstein*, pp.539–541. Penguin, London.

[26] Cotgreave, P. and Belsham, C. (1997) Home in the highlands. *Lifewatch*, Autumn 1997, pp.14–15.

[27] *Darwin Initiative for the Survival of Species: Fourth Report*, Annex A. Department of the Environment, Food and Rural Affairs, London (2001).

Cunningham, a veterinary pathologist who knows more about wildlife diseases than almost anyone in the world, with the local knowledge of the Parsee population. The Parsees have lived in Bombay since the 1660s, when parts of the area came under the control of the British, who operated a policy of toleration for all religions. If that sort of collaboration cannot make progress, we are all in trouble.

* * *

However clever ecological scientists are, they cannot avoid the fact that studying the environment sometimes involves temporarily damaging small parts of it. For this reason, many of us have mixed feelings about some kinds of environmental research.

The ambiguous nature of our interest in ecology could not be clearer than in the case of tuberculosis in badgers and domestic cattle. Bovine tuberculosis is a major problem on dairy farms in many parts of the world. It has serious implications for the health and welfare of infected cattle. Approximately 20% of all herds of cattle in England and Wales contain cows whose skin 'reacts' to the standard test for tuberculosis, and are thus judged either to have the disease, or to have harboured the bacteria at some time, and hence to pose a threat to the health of others. In Ireland, more than 30,000 cattle have to be destroyed every year because of the disease, at an annual cost of some £50 million.

Although this is not obviously an environmental issue, its ecological dimension becomes clear once we realize that the bacteria that cause the disease are also found in various species of wildlife, and that the solution to the problem may well lie in understanding the interactions between domestic cattle and wild animals.

In England, the bovine tuberculosis bacterium has been identified in at least ten different species of wild mammals, ranging from moles (small, underground insectivores) to red deer (large herbivores), and including rats (small generalists) and mink (middle-sized carnivores). Special attention has been paid to the badger, because some studies have shown it to be particularly susceptible to infection by tuberculosis bacteria, and because it is especially common in South West England where the incidence of tuberculosis in cattle is unusually high.[28]

An obvious inference is that badgers are responsible for spreading disease among cattle. As a result, cattle farmers are understandably keen to exterminate badgers from their land. According to Brian Jenkins of the National Farmers' Union, 'controlled culling [of badgers] is needed if the

[28] Krebs, J. (1997) *Bovine tuberculosis in cattle and badgers. Report to the Right Honorable Dr Jack Cunningham MP*, pp.13 and 26. Ministry of Agriculture, Fisheries and Food, London.

nation's cattle population is not to be placed under threat'.[29] But the badger is a protected species and naturalists, animal lovers and environmentalists are horrified that badgers should be killed. Many are especially upset because they think it unusually cruel to block the entrance to the animals' underground burrows and then gas them.

The protestors also make the case that nobody knows for certain whether badgers do indeed spread tuberculosis. They point to the fact that the disease is found in French cattle herds where it cannot be spread by badgers, because surveillance programmes have never identified tuberculosis in any French badgers. It seems unlikely that badgers caused two outbreaks of disease in cattle in Switzerland in 1996, because no infected badgers have been found there for more than twenty years. These sceptics have powerful allies. Tony Banks, an outspoken member of Parliament, who was a government minister between 1997 and 1999, made a forthright criticism of the government run by his own political party, saying that 'like a lot of things at the Ministry of Agriculture, killing badgers seems to be based more on voodoo than science'.[30]

In 1997, the government tried to break the impasse by setting up an independent scientific review, under the leadership of the distinguished ecologist Professor Sir John Krebs. Krebs's team reviewed every scrap of evidence, and concluded that although it was likely that badgers did spread bacteria among domestic cattle, it was not possible on current evidence to judge the scale of their contribution to the problem, or to make any assessment of the efficiency or cost-effectiveness of policies based on culling badgers.[31]

The government accepted the review group's suggestion that the only way to settle the issue once and for all was to carry out an experiment. Across the country, large areas of farmland would be identified and then divided into three equally sized sections. In one section, badgers would be exterminated as thoroughly and rigorously as possible. In the second section, badgers would only be killed if they were living near cattle known to be infected with the disease. The third section would act as an experimental control, in which no badgers would be killed, so that the researchers could tell whether the other areas showed fewer cases of disease than 'natural' areas. More than 200 staff were dedicated to carrying out the seven-year experiment, the costs of which were estimated at £34 million.[32]

[29] BBC News Online, 17 August 1998.

[30] *The Independent*, 13 April 2001.

[31] Krebs, J. (1997) *Bovine tuberculosis in cattle and badgers. Report to the Right Honorable Dr Jack Cunningham MP*. Ministry of Agriculture, Fisheries and Food, London.

[32] *The Times*, 22 January 2001.

Badgers were not to be gassed, but were to be trapped using humane traps (bated with peanuts) and then despatched instantaneously using a pistol.

Problems began to arise almost immediately. Animal lovers felt so strongly about the issue that some of them went around at night trying to release trapped badgers before government officials came along in the morning to shoot them.[33] At a demonstration in the popular and beautiful Peak District National Park, protesters dressed up in badger costumes and claimed that the public were 'horrified' by the experimental cull. Dr Elaine King of the National Federation of Badger Groups said that after the Park's twenty million visitors had gone home, 'the Ministry of Agriculture will be slaughtering one of the most popular animals in the countryside'. She called the affair a 'national disgrace'.[34] There were even accusations, vigorously denied, that the government was 'breaking its own laws' in carrying out the experiment.[35]

But the environmentalists and animal lovers were not the only ones complaining. In the control areas, farmers were unhappy that they were forbidden to kill badgers, which they were certain were spreading tuberculosis into their cattle herds. Krebs's committee had, after all, concluded that it was likely that some cattle did pick up the bacteria from badgers. People involved in the trial began to have strong suspicions that some landowners and farmers were killing badgers even though their farms were in 'control' areas. This was not only illegal, but was rendering the experiment useless.

In other words, it is proving almost impossible to conduct a relatively simple experiment, even though it was designed to solve a genuine ecological problem of real economic interest, and is concerned with the welfare of cattle and with restricting the spread of an extremely unpleasant disease. The reason, at least in part, is that we do not like the small-scale, short-term, ecological consequences and environmental effects of research that seeks to find long-term answers to questions about the ecosystem in which we live.

* * *

Even when the international political intricacies of paying for environmental science are sorted out, as they are in the case of the Darwin Initiative, and even when disagreements about the ethical nature of a particular experiment or project can be overcome, there remain inherent

[33] BBC News Online, 9 November 1999.

[34] BBC News Online, 6 August 2000.

[35] *The Times*, 25 November 2000.

difficulties in developing suitable methods for funding scientific research in the fields of ecology, the environment and biodiversity.

Two kinds of scientific research demonstrate some of these difficulties, both focusing on the fact that environmental science depends more heavily than some other disciplines on the incremental assembly of small and often unexciting facts and insights to build a larger picture. One of the underpinning disciplines is formed from long-term studies, where each year we collect only tiny fragments of a jigsaw, and where we cannot even imagine the bigger picture until we have a lifetime's data. The other is simply the describing, naming and cataloguing of the vast array of life forms with which we share the planet.

* * *

Long-term studies of ecosystems and organisms tend to generate more significant environmental insights than shorter-term ones. The only reason we know that the Earth's temperature is rising is that we have information about how hot the world was decades and even centuries ago. We know that the song thrush is declining in numbers, or that the level of pollution in the Thames is falling, because these things have been measured for decades.

Long-term ecological studies have provided an exceptional number of insights into the workings of the Earth's environment. Back in 1971, a group of ecologists began a study of the red deer population on the island of Rum off the west coast of Scotland. The group is led by Professor Tim Clutton-Brock, and it has provided a wealth of scientific insights into the dynamics of a population of wild animals.[36] After 30 years, Clutton-Brock and his team are still publishing new results in *Nature*, one of the world's most prestigious scientific journals.[37]

A similar study of the Soay sheep on the Atlantic archipelago of Saint Kilda has recently resulted in a discovery that sheds even greater light on the interconnectedness of climate change and biodiversity. The Soay sheep is an ancient breed that has lived on the Saint Kilda islands for centuries. Although standardized scientific counts of the sheep did not begin until the 1980s, there had been a recognition of the value of censuses thirty years earlier, so that by 2001, Tim Coulson and colleagues from the University of Cambridge were able to work with rough-and-ready data covering almost 50 years. They were able to work out what factors

[36] Clutton-Brock, T.H., Guinness, F.E. and Albon, S.D. (1982) *Red Deer: Behaviour and Ecology of Two Sexes*. University of Chicago Press, Chicago.

[37] Kruuk, L.E.B. et al. (1999) Population density affects sex ratio variation in red deer. *Nature*, Volume 399, pp.459–461.

affected the fortunes of the Soay sheep on the main island, which is called Hirta.

Among other factors, they found (not surprisingly) that the gales and heavy rain of the Atlantic winter can have a severe effect on the numbers of sheep, and that the age of an individual animal affected its chances of surviving a bad winter. According to Coulson, 'individuals older than six years are significantly more likely to die in years . . . when winter weather is bad, than adults in their prime'.[38] More importantly, the team discovered that the tendency of the sheep population to plummet in bad weather was correlated in particular with the North Atlantic Oscillation, or NAO.[39]

The NAO is a large-scale movement of atmospheric mass between the subtropics and the North polar region. At one extreme, there is a very high pressure in the subtropical region around the Azores and Spain, and a correspondingly low pressure around Iceland. In years when this occurs, there are more (and stronger) winter storms in the Atlantic, and warm, wet winters in Europe. At the other extreme, the difference in pressure between the Azores and Iceland is much less pronounced, and this reduced gradient in the air pressure leads to less stormy weather in the Atlantic Ocean, and brings cold air to northern Europe.

The oscillation moves between its two phases from time to time, but shows a tendency to stay in the same phase for several years in a row. In the past 30 years, the NAO has shown a tendency towards more and more extreme pressure differences between the subtropics and the polar region, leading to milder winters in Europe, increasingly violent storms in the Atlantic, and changes in the production of zooplankton and the distribution of fish. It is likely that these changes are caused at least in part by human activity, such as the production of greenhouse gases, and that they represent one manifestation of continuing global climate change.[40]

Only by having access to data collected over a 50-year period were Coulson and his colleagues able to uncover the potential importance of the North Atlantic Oscillation on populations of animals living in the Northern hemisphere. The Saint Kilda island wardens who began counting sheep in the 1950s could never have dreamt that their work would contribute to an understanding of how human-induced climate change

[38] Coulson, T. (2001) Age, sex and the weather: Modelling population dynamics among Soay sheep, pp.576–577 in Macdonald, D. (2001) *The New Encyclopedia of Mammals*. Oxford University Press, Oxford.

[39] Coulson, T. et al. (2001) Age, sex, density, winter weather, and population crashes in Soay sheep. *Science*, Volume 292, pp.1528–1531.

[40] Hurrell, J.W. (1995) Decdal trends in the North Atlantic Oscillation regional temperatures and precipitation. *Science*, Volume 269, pp.676–679.

might impact on the organisms that live in the Earth's ecosystems. Indeed, they cannot have had the slightest idea how important their efforts would turn out to be, but they started counting the sheep anyway.

It is clear that if we want to be confident in our beliefs and predictions about environmental change in the future, we need to start investigations now that will not produce useful and meaningful results for years to come. However, while they may have obvious uses, long-term studies present a problem for those who organize and pay for our scientific endeavours. The people who start these enduring projects may not see them come to fruition in their lifetimes, and it follows that the people who start to pay for them may not see any useful return before they die.

That ethos does not fit with the annual budgetary negotiations of governments. If a project failed to produce at least one significant finding every couple of years, it would be difficult to preserve its funding, regardless of the insights that might accrue over the long term. A lack of long-term funding is one of the principal concerns of many environmental scientists.

* * *

Perhaps the greatest problem for environmental conservationists is knowing what is out there to conserve. Because we share the planet with a vast diversity of different life-forms, we have not yet found the time to study them all in any detail. Indeed many have not yet been discovered.

Once a species has been discovered by science, it is a given a two-part scientific name, which is either Latin or Latinized-Greek. The Swedish naturalist Carl Linnaeus invented the system of giving organisms two scientific names, with the first name being common to all organisms in a particular genus of similar species, and the second name being unique to the species. Thus, the American robin is given the name *Turdus migratorius*, while the closely related European blackbird is called *Turdus merula*.

In 1758, Linnaeus gave names to about 4,400 species of animals and 7,700 plants and ever since then, taxonomists have added to the list.[41] Every year, thousands of previously unknown species of organisms are given scientific names and descriptions. A sample of just 50 from those that were new to science in 2000 shows that although there were five plants, two frogs, two fungi, a mammal, a bacterium, a protozoan, a fish and a snake, the majority were insects including two fleas, two wasps, six mayflies and 19 beetles.[42]

[41] Linnaeus, C. (1758) *Systema Naturae*. Tenth Edition.

[42] The sample comprises the first 50 extant species from publications revealed by a search for papers with the words 'new species' in the title on the Web of Science database, using information supplied by the Institute of Scientific Information.

It is hardly surprising that very few new bird species are discovered, because birds are easy to see and have already been well studied throughout much of the world, so our lists are already fairly comprehensive. New birds are occasionally found, but they are very unusual. When two students called Paul Salaman and Thomas Donegan discovered the chestnut-capped piha in the mountains of Colombia in the summer of 2001, it was sufficiently newsworthy to receive a 500-word article in *The Times*. It was, said Mr Donegan, the 'ultimate experience', even though the bird was small, grey and easy to overlook.[43] However, the rate at which new insects are still being discovered leads us to suppose there are still a great many novel species that have yet to be catalogued.

Describing new species is, in general, a worthy but unexciting job. Only occasionally is a new species considered so interesting that it deserves scientific and media interest. An undiscovered mammal in a well studied country, such as the bat species discovered in 1997 in Western Europe,[44] is certain to interest the newspapers,[45] while a tiny creature of a wholly unfamiliar type, unrelated to all known species,[46] was remarkable more for the fact that it was discovered living on a lobster's lip than for anything about its fascinating biology.[47]

But most new species are unremarkable beetles that seem to be rather similar to others that are already known, so they are not accorded such fame in the press, and the scientists who discover and describe them are never going to win a Nobel Prize. The work of cataloguing life – known as taxonomy – is not considered cutting-edge science, and the funding organizations are not keen to support it. Neither are they especially enthusiastic about funding systematics – the study of how the many species are related to one another – except where such studies can reveal new insights into the process of evolution or other biological processes.

In the UK, the distinction between the most exciting science and the routine of description is plain in the mechanisms for funding research. The Natural Environment Research Council, which is funded by the Office of Science and Technology, is not keen to use its precious resources on taxonomy or basic systematics, when there are plenty of cutting-edge investigations to fund. The main institution that carries out taxonomy is

[43] *The Times*, 20 August 2001.

[44] Barrat, E.M., Deaville, R., Burland, T.M., Bruford, M.W., Jones, G., Racey, P.A. and Wayne, R.K. (1997) DNA answers the call of pipistrelle bat species. *Nature*, Volume 387, pp.138–139.

[45] *The Guardian*, 9 May 1997, p.7, reported that the discovery had left scientists 'in shock', whereas the two scientists quoted in the article actually used the words 'astonishing' and 'fascinating'.

[46] Funch, P. and Kristensen, R.M. (1995) Cycliophora is a new phylum with affinities to entoprocta and ectoprocta. *Nature*, Volume 378, pp.711–714.

[47] *The Washington Post*, 14 December 1995; *The Times*, 14 December 1995.

the Natural History Museum which, as a museum, is part of a completely unrelated government ministry, the Department of Culture, Media and Sport. We can hardly be surprised that this ministry has not focused on the value of taxonomic and systematic research, given that its official remit is about 'cultural and sporting activities and . . . strengthening the creative industries'.[48]

Many modern biologists, whose interests vary from individual genes within the human body to the working of complex ecosystems like the Amazon rainforest, believe that taxonomy and systematics 'underpin research into the conservation of biodiversity, or crop cultivars as used in agriculture' and are important in relating what we now know about genetics to what we understand about ecology and evolution. But in the UK, the same people perceive that the study of systematics is 'under real threat' because ministries like the Department of Culture have, understandably, focused on other priorities, while the Research Councils have prioritized research that is more scientifically exciting.[49]

In 1992, the House of Lords carried out an inquiry into the area, and concluded that steps needed to be taken to support what was becoming an overlooked but important area of science.[50] The British government responded with funding to establish a Systematics Forum, but after a few years, the money ran out and there were no plans to continue the exercise. It became known as 'the second death of systematics'.[51] In 2002, the House of Lords was so disappointed, it decided to establish a second inquiry.[52]

If we care about the environment as much as we say we do, we must empower and instruct our governments to work together and to invest in research that will not always bring them headlines – investigations of viruses in vultures and descriptions of unspectacular insects. If we do not, the planet will change out of all recognition in the next 100 years, and we will not even know what we have lost.

[48] <www.culture.gov.uk>.

[49] *Science Policy Priorities 2001*, p.7. Institute of Biology, London (2001).

[50] *Systematic Biology Research*, First Report of the House of Lords Select Committee on Science and Technology, Session 1991–1992. [HL 22-I].

[51] *Research Fortnight*, 4 July 2001, pp.18–19.

[52] *House of Lords Select Committees, Weekly Agenda*, No 15 of Session 2001–2002, p.11. Stationery Office, London (2002).

5 Hollywood movies, atom bombs and the defence of the realm

Developments in defence research during and after the Second World War demonstrate how scientific progress depends on networks of communication and collaboration

THE Second World War happened at exactly the moment in history when the film industry had truly come of age. Hollywood was ravenous for exciting material to turn into movies. It did not matter that the conflict had been brutal and ugly, that people's sons and daughters had been massacred in their hundreds of thousands, or that distrustful nations were trying to rebuild their relationships with one another, after those relationships had been savagely fractured, smashed and disfigured by six years of the bitterest and most shaming hatred the human race has ever seen.

Hollywood had a job to do, and for more than a quarter of a century after the end of the war, film-makers produced a string of classic movies that used the Second World War as a backdrop. Some of those films were about individual wartime battles and events, like *The Dambusters* (made in 1954), *The Great Escape* (1964), and *The Battle of Britain* (1969). Other films, such as *The Sound of Music* (made in 1965), told stories with timeless themes, but used the absurdity and uniqueness of wartime as a convenient device to set the scene. After a couple of decades, the television industry decided that the time was ripe to start bringing humour to bear on the subject of world wars, and situation comedies such as *Dad's Army* and *'Allo 'Allo* became popular.

For many of us in the world today, these films and television shows are the most frequent and important source of information about the War, and it is not always easy to be aware that when some of these films were first shown, the events they portrayed were fresh, perhaps even raw, in some people's minds. Although many of these movies were not made until the 1960s, some of the real people they sought to portray – young prisoners of war in Colditz Castle, or Austrian children born in the 1930s – were barely middle-aged when they watched themselves depicted on the silver screen.

Oblivious to the concept that these people and their memories were still very much alive, many of us grew up seeing the events in the films as history, being part of some great expanse of time called 'the past'. We thought that the Battle of the Bulge and the Dunkirk evacuations were barely more immediate than Henry the Fifth's siege of Harfleur, the defeat of the Spanish Armada, or the bravery of the South Wales Borderers in the face of overwhelming hordes of Zulu warriors at Rorke's Drift.

In part, this misunderstanding is based on the fact that the real essence of a story, portrayed by a good film, is largely unaffected by the precise circumstances in which the film's action takes place. Even the century is unimportant. *Zulu* is about a small group of men showing unusual fortitude in the face of crushing odds, and so is *The Great Escape*. *The Dambusters* is about one man's obsessive crusade to make other people see things his way, and so is *Henry V*. Their clothes were different, their speech differed, their enemies and causes were different, but if a time machine ever allowed Henry of Monmouth and Sir Barnes Wallis to meet, they would almost certainly find that they had much in common.

* * *

One way to understand the true historical context of old films is to appreciate the technological efficiency of the winners – King Henry used fine rhetoric and a sword, Barnes Wallis used fine rhetoric and a bouncing bomb. It is still unreal and almost meaningless for many of us to know that our grandfathers were actually too old to fight in the Second World War, or that our parents were evacuated from the cities, or had evacuees billeted on them in the countryside, or travelled half way round the world to fight.

What we can understand, however, is that Barnes Wallis broke the dam at Mohne by studying the physics of a solid ball bouncing on water, and that Michael Caine's character in *Zulu* had guns and other military hardware not possessed by the Zulu army. He would not have told the private soldiers under his command to 'wait until you see the whites of their eyes' before shooting if the thousands of warriors in King Cetewayo's army had all been issued with rifles.

In *The Battle of Britain*, the character of Hermann Goering stands on the northern French coast and asks his senior airforce officers what they need. He has just finished berating them for allowing cowardice among the fighter pilots, whose job is to protect the Luftwaffe's bombers in the air. By way of conciliation after his outburst, he implies that they can have anything they want if it will help them beat the British Royal Air Force. Taking Air Marshal Goering's remarks at face value, one of the officers makes the mistake of asking for a squadron of Spitfires to replace the wretched and substandard aircraft possessed by the Luftwaffe. The look on Goering's face shows that he is not impressed, because he knows he cannot deliver aircraft of such technological quality. His expression leaves no doubt that the career of the bold airforce officer is unlikely to progress well afterwards.

Of course, this was a film made from the perspective of the victors, and the scene can hardly be treated as an historical documentary, but its message is broadly sound. Goering's reputation faded after 1940 not because he was an incompetent leader but, at least in part, because better technology allowed the allies to win the battle for air supremacy.[1]

* * *

There are in essence just three ways to win a military battle. One is to have an overwhelming number of troops, so that, however many are lost, there will be waves more to grind down the enemy. This is partly how the Roman army operated in its heyday, shipping huge numbers of foot soldiers around the Empire to crush rebellion and conquer new races. It was exploited in the opening scene of the film *Gladiator* (2000), which is a work of art, depicting both the numerical strength and the discipline of the Roman legions. As the German barbarians shake their primitive axes and scream obscenities, a friend of General Maximus observes that 'people should know when they are conquered'.

The second way to triumph in a military fight is to have a brilliant and bold tactician who takes incredible risks but who can motivate the poor bloody infantry into believing that the most unlikely of victories is possible. Hannibal, the Carthaginian general of the third century BC, knew that this kind of vision and courage could overcome a much larger force. With a force of around 50,000 men at the Battle of Cannae in 216 BC, Hannibal managed to surround a force of Roman soldiers much greater in number than his own, and crushed the army of Quintus Fabius. Almost all the men of Hannibal's Carthaginian army survived (only about 6,000 died), but fewer than a fifth of Fabius's 80,000 Romans escaped. A further 10,000 were captured alive, but the vast majority – almost 60,000 of them – were slaughtered.[2]

The third, and surest, way of defeating an enemy by force is to deploy superior technology. It is the premise of almost every science fiction film ever made that the alien overlords use scientific skill and gadgetry not available to whichever miserable race ends up being conquered. It was the same premise that allowed a Western Alliance dominated by Britain and America to win the Gulf War against Iraq so quickly in 1991, and then to

[1] Magnusson, M, (ed.) (1990) *Chambers Biographical Dictionary*, p.600. Chambers, Edinburgh.

[2] Isaacs, A., Martin, E.A. et al. (1998) *Oxford World Encylopedia*, p.247. Oxford University Press, Oxford.

demonstrate military superiority through air raids in the Balkans in 1999, almost without the loss of a single Western life.

* * *

One of the best known examples of the strength of military technology comes from the resounding English victories at Crecy and Agincourt during the Hundred Years War with France. In both cases, the odds were massively in favour of the French, and Shakespeare's famous account of Agincourt implies that these odds were overcome by the enthusiasm and leadership of Henry V, coupled with the God-given righteousness of his cause. But the truth is that superior technology was just as important in winning the day.

The English longbow was a frighteningly efficient killer in the hands of Henry's skilled archers. They had trained hard, and that training paid off because when they used the bow skillfully, they had a normal effective range of 200 metres. In experienced hands, a longbow could be used swiftly and effectively over a distance of at least 300 metres. When an archery enthusiast, Sir Ralph Payne-Gallwey, tried his hand at firing a 400-year-old longbow in 1901, he managed to shoot an arrow almost 400 metres across the Menai Strait in North Wales.

It is doubtful whether medieval archers could achieve such distances on the battlefields of France, mostly because of the speed with which they were showering the enemy with arrows. Each archer was shooting ten or more arrows every minute. Robert Hardy has estimated that something like half a million arrows were shot by the English archers during the few hours of battle at Crecy, where the hail of arrows was said to be 'so thick it seemed liked snow'. At Agincourt, the French cavalry were 'quickly compelled amidst showers of arrows to retreat and fly'.[3]

A bolt from a longbow could penetrate metal chain mail at 100 metres. The skill and strength involved was considerable – it might require a force of almost 50 kg to draw a bow. The longbow was typically made from the supple wood of a yew tree, and the arrows they fired were almost a metre in length. It was six centuries before a rifle was invented that was able to fire bullets at the same speed as a longbow's arrow.[4]

The odds at Agincourt were crushingly in favour of the French King, who may have had ten times as many men in his army as there were in Henry's force. The English army consisted of a few thousand, while the French had at least 40,000 men, and probably more. But when the chap-

[3] Hardy, R. (1986) *Longbow: A Social and Military History*. Mary Rose Trust, Portsmouth.

[4] *Encyclopaedia Britannica*, article entitled 'Longbow'.

lain of the English forces noticed that 'the enemy's crossbowmen retreated from the fear of our bows', he was not mistaken. At the end of the day, 10,000 Frenchmen had been killed, including half the country's noblemen, while the English losses may have been fewer than 100, and were certainly not more than 500. English and Welsh archers were extremely lucky to have had superior technology on their side.

The most remarkable thing about the French defeat at Agincourt is that the winning technology was not only known to the French army, but was actually possessed by them. After the battle, many longbows and arrows were discovered, but they had not even been unpacked from the bales into which they had been tied for transportation to the field. Despite having the technology, the French persisted in using cumbersome crossbows, even though they had been shown to be inferior battle weapons at Morlaix seventy years earlier, when a horde of Genoese crossbowmen in the pay of the French had been shot to pieces by Edward I's longbow archers.[5]

* * *

The Second World War, beloved of the Hollywood film-makers, saw some of the most significant developments in military technology the world has ever experienced. Among these, radar stands out as one of the most interesting from a scientific point of view. The radar effect had been known since the 1920s, and was 'rediscovered' in 1930 at the Naval Research Laboratory in Washington, DC, when L.A. Hyland observed that an aircraft caused a fluctuation in a radio signal when it flew through the signal beam. The British picked up on the technology and, by 1938, the first British radar system, known as the Chain Home, went into 24-hour operation and remained fully operational until after the end of the War in 1945.

In fact, although Britain was using radar as early as 1938, at the outbreak of war in 1939 it was Germany that had progressed further than any other country in developing the technology. By 1940, however, Hitler believed the War was almost over, and rather than use precious resources on something that probably would not be needed, his military command halted further work on radar.[6]

Winston Churchill allowed British engineers to press ahead with their research, and in about 1940, they invented the cavity magnetron, which allowed the first use of high-powered microwaves. This was a powerful development and early in 1941, radar was successfully used to detect German bomber aircraft which could then be intercepted by allied fighter

[5] Hardy, R. (1986) *Longbow: A Social and Military History*. Mary Rose Trust, Portsmouth.
[6] *Encyclopaedia Britannica*, article entitled 'Radar'.

planes, controlled from a base on the ground. By knowing exactly where the German bombers were, the limited number of fighters could be deployed to best effect.[7]

The British shared their new knowledge with the USA (which had not then joined the War), and it was the flow of information between the various parties that led to the significant breakthroughs. The Welsh engineer Eddie Bowen later said of the British team that 'in many ways it was more like a scientific convention than a research laboratory, except that it was a convention which kept running year after year'.[8] Sharing information with the Americans paid off. It was the Massachusetts Institute of Technology that went on to develop about 150 different radar systems in the following five years. Of these, one of the most important was a system for gunfire control, in which anti-aircraft guns could be aimed on target without the need for searchlights. It caught the Germans completely unprepared and made a significant contribution to the allied victory.[9]

However, the most dramatic example of the efficacy of scientific prowess in winning wars comes from the atomic bombs that fell on Japan at the end of the Second World War. The first bomb, dropped on Hiroshima from an American B-29 bomber, killed upwards of 70,000 people immediately. The bombing took place on 6 August 1945, and used uranium to create an explosion with a force of more than 15,000 tons of TNT. Three days later, a plutonium bomb was dropped on Nagasaki, and a further 40,000 people lost their lives. Japan immediately capitulated, and the war in Asia was over.

Events surrounding the development of nuclear weapons are explored in Michael Frayn's play, *Copenhagen*. He examines the motives of the physicists Niels Bohr and Werner Heisenberg, and charts the changing nature of their relationship, basing his story around the mystery of a meeting that took place in Copenhagen in 1941, at which the two men argued, but which has never been fully explained. The whole atmosphere is clouded by the fact that Bohr, a Dane, lives in a country that has been occupied by Heisenberg's country. The developing plot explores the morality of whether it could ever be right knowingly to develop a weapon of mass destruction.

For the audience, watching events unfold between the two men is gripping, at times confusing, and certainly thought provoking. But we

[7] Latham, C. and Stobbs, A. (1999) *Pioneers of Radar*, p.52. Sutton Publishing, Stround.

[8] Quoted in Buderi, R. (1997) *The Invention that Changed the World: How a Small Group of Radar Pioneers won the Second World War and Launched a Technological Revolution*, p.98. Simon and Schuster, New York, NY.

[9] *Encyclopaedia Britannica*, article entitled 'Radar'.

watch in the knowledge that atomic weapons were indeed developed, and that in the end it was Bohr and not Heisenberg who contributed most directly to the progress of the relevant research. Most of all, we watch their conversation in the knowledge that these horrific devices were actually used to kill tens of thousands of fellow humans. The atomic bomb is one, enormous, reminder of the power of science and technology, but the fact is that there are thousands of examples of military technology winning battles, winning wars, and changing the course of history. We do not need to make moral judgements about military history to recognize that technology helps win wars.

* * *

Even in modern wars, except in a few cases, most of us do not really know enough of the historical context to have any real idea of who was in the right and who were the bad guys. But we know enough of human nature to know that in at least some cases, there is no clear right or wrong. That is why some conflicts, like that between Israel and the Palestinians, are so intractable. In these cases, fighting happens because it has become traditional and ingrained, or because nobody can think of a better way forward. Each side may have a legitimate cause for complaint against the other but their desires are, or seem to be, completely incompatible.

We might examine the role of the longbow in the Hundred Years War or the atom bomb at the end of World War II not to glorify fighting, but merely to demonstrate that science and technology matter in terms of defence and military offence. We do not have to like the idea of fighting to realize that history has been changed by scientific advances in the armed forces.

Sometimes, that historical change has come in large-scale political consequences, such as the successful fight for independence in eighteenth-century America. At other times, it has been manifest in the form of dramatic effects on the lives of individual human beings. For the individual French crossbowman at Agincourt or Crecy who lost his life, and for his orphaned children, the longbow, skillfully wrought by an English yeoman from the bough of the yew tree, was an instrument of life-wrenching devastation. Who can tell what the event meant to the English bowman himself? We know the names of many of the English and Welsh archers at Agincourt – men like Jenkin Fustor, Evan ap Griffith and David Whitchurch – although we know precious little else about them.[10] Perhaps, many years later, when they were old men and England herself was in turmoil, they wondered whether it had all been worth it. Or

[10] Hardy, R. (1986) *Longbow: A Social and Military History*. Mary Rose Trust, Portsmouth.

perhaps they liked to tell their grandchildren about their youthful exploits in France.

The effects of wars on the broad sweep of history, and the effects of battles on individual lives, mean that is foolish to ignore the science of military defence. That is why defence research and development should always be relatively high on the political agenda, and why the public interest demands that our political leaders structure our military research in ways that minimize the threats that the peoples of the world face from military offence. Indeed, for forty years it was argued that technology prevented war between the world's superpowers. Because both the USA and the Soviet Union invested in military technology, neither had any kind of superiority over the other, and there was never any point in either one attacking the other.

* * *

The end of the Cold War saw a new phenomenon in the politics of many countries, especially many Western political establishments. After the fall of the Berlin Wall, many political leaders spoke avariciously of the money they could save from the 'peace dividend'. In the new, safer, world that was dawning, there would be no need to invest such a high proportion of the nation's wealth in the armed forces. The public interest would be served by saving money, and using it either to give citizens a cut in taxes or to improve more popular public services, such as health and education.

In the UK, the savings fell disproportionately heavily on the Ministry of Defence's programme of scientific research. Between 1994 and 2000, the Ministry's research budget fell by 27%, a saving equivalent to almost £200 million every year.[11] The same trend occurred in America. The federal research budget for the Department of Defense fell by almost 4% in a single year between 1996 and 1997. Overall, it fell from almost 21% of the federal research budget in 1990 to just 13% in 1998.[12] In Britain, this cut in funding led ultimately to changes in policy that have the potential to cause great harm to the public interest. They threaten to weaken the nation's scientific and defence capabilities, not just because we now spend less money, but also because the drive to save money has led to a reorganization that appears to ignore some of the fundamental features of the way science works.

* * *

[11] *SET Statistics 2000: A handbook of science, engineering and technology indicators*. The Stationery Office, London (2000) [Cm 4902].

[12] *Science and Engineering Indicators*, pp.4.21–4.22. National Science Foundation, Washington, DC (1998).

Science advances partly because of brilliant people having clever new ideas, and partly because those ideas develop in a series of discussions, sometimes held face-to-face, sometimes held in print, sometimes even held across generations. In other words, there are great leaps forward, punctuated by periods of more modest, incremental progress.

The incremental progress depends entirely on the discussions that occur at conferences, in the pages of journals, in the bar and over coffee. The consequences of bouncing ideas around at lunch-time are portrayed in the film *Enigma* (2001), in which it seems that the two main protagonists always have their best ideas during a conversation over a meal, or while two people are chatting as they walk in Bletchley Park, and not while sitting at their desks, trying to have clever thoughts.

Many of the breakthroughs in wartime radar came precisely because the English shared their results with the Massachusetts Institute of Technology, a transatlantic dialogue on a larger scale, but similar in nature to the dialogue between the two main characters in *Enigma*. The breakthrough in atomic weaponry that ended the War in the Far East came when a disparate group of scientists were thrown together by the Allies, and engaged in the same sort of conversation.

Apparently disparate scientific issues turned out to be deeply interlinked and, consequently, there was a need for a two-way flow of information between the various different people involved in discovery, and also between those researchers who discovered new things and the people whose job was to use the knowledge and understanding that the scientists had generated. That is why the wartime radar expert Eddie Bowen talked enthusiastically about an ongoing seminar, in which everyone shared their thoughts and results. In scientific and technological progress, anything that creates barriers to this communication is likely to create significant problems, which will hinder both the research itself and the development of new applications.

* * *

It is because scientific conversations are so important that many people opposed the partial privatization of the UK's Defence Evaluation and Research Agency. Surprisingly, the main issue was not the apparent inappropriateness of throwing part of the defence of the realm to the mercy of the marketplace. In fact, there were perfectly defensible reasons for allowing private enterprise the chance to run some of the Agency's activities.

The opposition to privatization sprang from a scientific point of view, not from an economic, strategic, or ideological one. The significant problem was that by privatizing part of the Agency, and retaining part in the public sector, the new arrangements were always likely to cause a decrease in the rate at which information flowed between the new private institution and the rump of the public agency.

The argument in favour of the scheme was that the Agency had two different roles. The most obvious was to carry out the research needed by the armed forces in order to keep them at the forefront of modern military techniques. This might include testing new guns and tanks, inventing new machines and gadgets, and developing techniques to predict the key information, such as weather forecasts, that is needed by military commanders in the field. Since this is such a sensitive and crucial role, everyone but the most rabid proponent of free market economics would agree that it could not be farmed out to the private sector. National security is the most essential of all public services, and its costs must be borne by taxpayers.

Moreover, all countries rely to some extent on the sharing of information with friendly nations, and those allies are unlikely to feel so strongly inclined to part with military secrets if they know that they are being given to a profit-making company in the private sector. This is a particularly important consideration for a small country like the UK, which depends on the goodwill of much larger international partners, especially the USA. Thus, strategic considerations demanded that the government retain ownership and control of the basic research and advice functions of the scientific arm of the defence establishment.

The other role of almost every research organization in the modern world is to generate wealth by commercializing new inventions. The complication for an organization like the Defence Agency is that many of its inventions will have uses that are of economic value in the civilian arena, and are not covered by any security considerations. Given that its job is to assist in the nation's defence, it is hardly appropriate for the Agency to use precious resources in the development phase of an invention that may or may not eventually reach the stage where it can be marketed and pay dividends in civilian life.

History teaches us, however, that it would be foolish to ignore the commercial potential of scientific discoveries that have spun out from the defence arena. The world trade in carbon fibres is now worth billions of dollars, with interesting uses in sports equipment, transport and construction. The tension experienced by fishing rods, skis and tennis racquets means that, ideally, they must be very strong but, for obvious reasons, sportsmen and sportswomen cannot excel if such equipment is too heavy. The desire for a material that is both lightweight and strong is shared by the makers of aircraft, where strength is a matter of life and death, and weight is a matter of great economic concern because heavier planes use more fuel then light ones.

The demand for light, strong materials has always been difficult to meet, but after carbon fibres were first made in the early 1960s by stretching acrylic fibres and heating them to carbonize the material, a superb solution was available.[13] The technology has improved tennis racquets and aircraft parts alike, but it was invented by the defence establishment.[14] One of the original reasons for making the fibres was that carbon absorbs poisonous gases. Many people believe that eating a small piece of charcoal is a good way of settling some kinds of stomach upset, and the reason is because carbon is generally efficient at absorbing noxious gases, which is also why it is used in deodorizing insoles for the shoes of people with smelly feet. On the same principle, carbon fibres are still used in the production of protective underwear for military personnel.[15]

Better tennis racquets, bendy fishing rods and devices for reducing the problems of malodorous feet are not, however, at the top of the list of priorities for support by the public defence budget. Private industry should risk its capital on such ideas, as it does in any other field of development, and should properly reap the rewards of any successes and the financial loss accompanying any failures.

In a perfect world, it might be desirable for the public purse to take the risk of development. Provided it did so on the same terms as a successful business, it might generate income for the nation and relieve the burden of taxes. In fact, at the same time as the British government cut its investment in defence research, it increased its investment in the development of related products. Being realistic, however, this kind of investment is unlikely ever to happen on the kind of scale that would allow the vigorous pursuit of a wide range of commercial possibilities. Thus, it is highly desirable that private finance should be used for the commercial development of civilian applications of defence-related research. That is why the British government decided to privatize part of its Defence Evaluation and Research Agency.

Ministers recognized that, in theory, there was a conflict between the needs of the two branches of the Agency. The scientists charged with

[13] Isaacs, A., Martin, E.A. et al. (1998) *Oxford World Encylopedia*, p.254. Oxford University Press, Oxford.

[14] In a debate in the House of Lords in 2000, Lord Hogg of Cumbernauld chose carbon fibres, liquid crystals and voice-recognition software as three of the most important examples of useful technologies that were first developed by the British defence research agencies. *Hansard* [House of Lords] 17 April 2000, column 531.

[15] Crystal, D. (1997) *The Cambridge Encyclopedia* Third Edition, p.210. Cambridge University Press, Cambridge.

advising the military top-brass on how best to defend the nation would be compromised if their funding were tainted by business interests, and those charged with generating income from new and exciting inventions would always be starved of cash, and could never really succeed, without industrial funding. So, under the new arrangements, the core activity of researching and advising on military security remains in the public sector, while the business of developing and marketing inventions has been hived off into what is, in effect, a new commercial company.

Unfortunately, such a scheme is bound to fail in the long term, because it fails to appreciate the way in which science moves forward, largely through serendipitous conversations in the tea room or bar. It will inevitably lead to a decrease in the amount and intensity of communication between the two new structures, relative to what occurred when they were two parts of the same organization. This is the typical failure of many schemes aimed at increasing 'efficiency', which place a financial value on everything, except those things to which it is impossible to give a realistic valuation. Social intercourse is a valuable part of any scientific enterprise and, in a world where 'efficiency' is everything, accountants charged with defending the public interest should realize this, even if we are all forced to admit that we have no justifiable way of coming up with a meaningful estimate of how much it is worth.

The effects of this ignorance about the process of scientific dialogue will be particularly strong in the defence world, because considerations about national security will mean that the scientists will not be able to circumvent the stupidity of the system by talking to one another in their spare time. The whole set-up could, ultimately, have the most devastating consequences for the effectiveness both of the military advisers who remain in the public sector, and of the scientists and technologists who will work in the private sector.

It is partly because they were involved in, or knew about, exciting scientific projects that military science advisers were in the past able to give the most up-to-date and expert scientific advice. Once they are starved of substantive contact with their friends and colleagues in the world of scientific development, they will rapidly lose their value. Likewise, those researchers who remain in the public sector, working on problems that have immediate and obvious defence implications, will be hampered. For security reasons, they will not be able to tell the private-sector technologists what they are doing, nor will they be able to ask what the technologists are up to, because the technologists will be gagged by the commercial considerations that govern institutions in the private sector.

All in all, the lack of effective communication will become universal, with foreign powers reluctant to share information with the nation, with the scientific developers reluctant to share information with the public sector scientists, and with the public sector scientists at a loss to know where to turn for stimulating and effective scientific discussion.

* * *

At first sight, opposition to the partial privatization scheme has the appearance of an ideological stance. Privatization was one of the great battles of British politics in the 1980s and 1990s, with the nationalized monopolies of power supply and the railways being sold off to the private sector. The opponents of state ownership were so comprehensively beaten that, in 2000, even the apparently left-of-centre New Labour government adopted a policy of making the UK the only country in the world to have a privatized system of air traffic control.

But in the case of the Defence Evaluation and Research Agency (DERA), opposing privatization has nothing to do with political leanings; it is based on the understanding that science can only operate to its full potential when barriers to communication are minimized. Criticising the partial privatization of the DERA does not solve the paradox of how the government can continue to get best value for taxpayers' money in terms of military research and advice, while leaving the risks (and rewards) of commercial development to the bankers and industrialists who understand them.

The solution most likely to succeed would be the creation of a publicly owned independent corporation, rather like the British Broadcasting Corporation, which is regulated by Parliament but which is allowed sensible commercial freedoms. This means the public can demand a certain level of service and, in the case of the BBC, the corporation is able to raise money from private sources where it is appropriate. The British Post Office has also been run along similar lines.

It would be easy to point out that nobody who knows anything about either the BBC or the Post Office could pretend that their administration has led to perfection, but such an observation would be facile and would entirely miss the point about the scientific value of defence research. It is the struggle for perfection, embodied in the desire to remove all 'inefficiency', which is the death knell of any long-term science policy.

Science is a messy business, and cannot be predicted. Its developments can come from anywhere, including a chance remark over a beer or a cup of coffee, and there is no point in pretending otherwise. Such an argument is not a charter for scientific laziness, or for giving researchers an unfettered free hand to waste time and money. It is a recognition of a truth that is universal and timeless – genius cannot be managed.

Within months of the privatization of DERA, problems had arisen. Simon Thomas, the Member of Parliament for Aberporth, was complaining to the government that the newly created company, QinetiQ, was threatening to axe an apprenticeship scheme that had previously trained many of each successive generation of aerospace experts.[16] His immediate concern was for the jobs that would be lost in his constituency, but there were deeper concerns that, by becoming semi-detached from other parts of the scientific community, the new company's actions were causing problems, the effects of which would not be felt for many years. 'Eventually', it was concluded, 'Wales will find that it does not have the next generation of experts, and the economy will suffer as the aircraft industry fails to thrive'.[17]

But the company itself was already protesting about not being kept informed about other developments. In October 2001 QinetiQ was reported to be 'concerned about being left of out consultations' on plans to increase links between research and industry put forward by the part of the old DERA that had remained in the public sector within the Ministry of Defence. While Professor Keith O'Nions, the Chief Scientific Adviser at the Ministry, said that he was considering 'the potential impact [of his proposals] on QinetiQ', John Chisholm, the Chief Executive of the firm, said that having been 'expelled' from the Ministry of Defence, his company thought the episode had been 'unfortunate'. A trades union representative, Sean Clarke, went further, 'People keep using the word "unfortunate", but it is absolutely crazy'.[18]

Defence research in America is treated quite differently. The Department of Defense says that it aims to be the best because 'nothing less is acceptable to us, or to the American people'.[19] It appreciates that science and engineering are a crucial part of this objective and spends US$38 billion a year on research and development, although this has fallen steadily as the 'peace dividend' has reduced its overall budget.[20] It has a vast network of agencies, laboratories and centres of research, such as the Space Vehicles Directorate, the Institutes of Surgical Research and the Navy Clothing and Textile Facility.[21] These facilities conduct fundamental, basic, strategic and applied research in a huge variety of fields, all

[16] *The Western Mail*, 19 September 2001.

[17] *The Western Mail*, 1 October 2001.

[18] *Research Fortnight*, 24 October 2001.

[19] <www.defenselink.mil>.

[20] Fossum, D. et al. (2000) *Discovery and Innovation: Federal Research and Development Activities in the Fifty States, District of Columbia, and Puerto Rico*. Rand, Santa Monica, California.

[21] <www.scitechweb.com/inhousereport>.

of it in the public sector. The 'central research and development organization' is the Defense Advanced Research Projects Agency, which 'pursues research and technology where risk and payoff are both very high and where success may provide dramatic advances'.[22]

This vast research machine spends federal funds in almost all of the 50 states, but also appreciates the value of closer ties with the public sector. For example, the Small Business Innovation Research Program aims to work with 'small technology companies' on technologies with a potential application in defence. Under the scheme, Savi Technology Inc. of California was given three grants to develop a radio tag that allows the journey of cargo ships to be tracked. The Department of Defense obviously thought it had value for money, announcing that it had contributed 'just US$2.5 million', and was equally pleased that the device would also have 'major applications in the private sector'.

The Department is also keen on ensuring that its own research results can be used effectively outside the defence arena. Its Office of Technology Transition exists both to ensure that advanced technologies are used for weapons systems, and also 'to assist in the commercialization of defence technology'. It has ensured that a radio antenna developed by Todd Kastle of the Air Force Research Laboratory can be applied both in the military and the civilian fields, saving about 90% of the costs of previous similar devices.[23] All of these activities add up to an impressive array of research and its application, all in the public sector, not because the Americans have a great love of state funding and state intervention, but because it is the only way of ensuring close links between all areas of relevant science, including the sensitive defence research and scientific advice that must stay in the public sector.

When, in late 2001, Delores M. Etter published a report on the in-house scientific activities of the Department of Defense, she used her position as Deputy Under Secretary of Defense for Research and Engineering to emphasize this point. The whole purpose of the report was 'to encourage scientists and engineers to communicate with their counterparts at other labs on problems of common interest'.[24] In the UK, this is no longer possible, because a group of politicians with an ideological belief in the superior efficiency of the private sector took no account of how science is actually done. Eddie Bowen's 'scientific convention which kept running for years and years' has come to an end.

[22] <www.darpa.mil>.

[23] <www.dtic.mil>.

[24] *Department of Defense In-House RDT&E Activities, FY2000 Management Analysis Report*, p.1. Department of Defense, 2001.

6 Eighteenth-century government and mad cow disease

Problems surrounding food and agriculture, especially those surrounding mad cow disease, prove just how important it is to place science at the heart of modern political decision-making

WE seem to have been particularly concerned about the safety of our food in recent years. The media has reported many stories about the health problems associated with the production and preparation of meat, milk and crops. No doubt journalists are playing on the fact that we are particularly susceptible to scaremongering about food. Because we must all eat, and because in the modern world we seem to have relatively little control over what goes into our food, we fear the hidden dangers of bacteria and poisons in the products that are demonized as salt-rich, high-caffeine, full-fat and 'artificial'.

Many people in farming communities believe that a great deal of fear is engendered because the urban majority no longer understands the rural way of life. The Countryside Alliance, which claims that half a million British people support its ideals,[1] says that rural communities feel 'discontent at the lack of respect shown by politicians for their needs and priorities'.[2]

Nevertheless, the people of Britain have built up the impression that they have been particularly badly affected by such scares. Edwina Currie, a junior minister for health, had to resign in December 1988 because of unguarded comments about *Salmonella* bacteria in eggs. In 1996, twenty people died in Lanarkshire in Scotland after an outbreak of a deadly form of the bacterium *E. coli*. *Listeria* bacteria in cheese, possible connections between tuberculosis in cattle and people, and potential health risks of eating genetically modified foods, have all hit the headlines.

The social and economic effects of such stories are very significant. For example, there has been a tremendous increase in the popularity of

[1] According to a New Release dated 18 May 2001, the Alliance estimates that half a million people would have joined its countryside march in London if it had not been cancelled because an epidemic of foot-and-mouth disease made extensive rural travel problematic.

[2] Countryside Alliance News Release, 16 February 2001.

organic foods, which are vigorously marketed as being safer than conventionally-grown crops and meat. In consequence, the amount of British farmland that is under organic cultivation rose dramatically, from around 50,000 hectares in 1996 to more than ten times as much in 2001, even though there is little or no hard evidence that organic food is safer than conventionally produced foods. The Food Standards Agency regularly assesses any available evidence on food safety, and its chairman Professor Sir John Krebs has said that people buying organic food 'are not getting value for money . . . if they think they are buying food with . . . extra safety'.[3]

One of the most well reported food scares in recent years was 'mad cow disease' which is portrayed as a peculiarly British disease, and which certainly appears to have originated in the UK. Known to scientists as BSE (short for Bovine Spongiform Encephalopathy), it appears that David Brown, the late agriculture editor of the *Daily Telegraph* gave the disease the name 'mad cow disease' in the late 1980s, and it is by that name that the disease became commonly known throughout the press and broadcast media.[4] Chapter 7 will explore how the Ministry of Agriculture mishandled the funding of relevant research.

Almost certainly caused by an unusual kind of protein called a prion, the disease affects the central nervous system, and the overwhelming majority of cases appear to be contracted through infected food. It is believed to have been spread through the cattle population when cows were fed foodstuffs that had been made with meat, including material from the central nervous systems of other animals. Similar illnesses have been observed in antelopes in zoos, which were fed the same proprietary cattle feeds, and in cheetahs and tigers that were probably fed infected carcasses. A human form of the disease, known as variant Creutzfeld Jakob Disease, or vCJD, has killed around a hundred people who are believed to have contracted the illness by eating foodstuffs containing the rogue protein.

The most remarkable thing about the tragedy of BSE is not the apparent incompetence or complacency of any individual or organization, but the way in which governmental structures hindered the good intentions of good people. It seems astonishing that the effects of a disease that was unknown until 1986 should have been exacerbated by the systems and geography of government that were created at least 250 years earlier. To understand how, it is necessary to examine a few of the quirks of modern British government, starting with a piece of outdated nomenclature.

* * *

[3] *Nature*, Volume 412, p.666.

[4] *Daily Telegraph*, 14 August 2001, p.23.

When, in 1992, Michael Heseltine became the cabinet minister in the British government with responsibility for trade and industry, he decided to revive a title that had not been used for many years – he dispensed with the title of 'Secretary of State for Trade and Industry' and instead called himself 'President of the Board of Trade'. The newspapers ridiculed him, suggesting that, having failed in his burning desire to become Prime Minister, he had found a way of being a President instead.[5]

Behind the jokes, Mr Heseltine's desire to be known as President revealed a fundamental feature of the way in which Britain is governed. The structure of government – its organization, its hierarchies, even its geography – are largely determined by factors that were relevant in the period between about 1550 and 1900; some of those factors are still relevant, but others are not. A great many of them date from the reorganization of government that occurred when Sir Robert Walpole became the first 'modern' Prime Minister in 1721.

There was little point in Mr Heseltine being President of the Board of Trade because the Board no longer existed. If it did, its membership would include the Archbishop of Canterbury. Even when it was formally in existence (from 1786 onwards), its meetings were entirely nominal, with none of the members attending except the President and Vice President.[6]

Other archaic titles are retained for other ministers, like the Chancellor of the Duchy of Lancaster and the Lord Keeper of the Seal. The chief finance minister, the Chancellor of the Exchequer, derives his title from a chequered tablecloth – the Exchequer – used by medieval tax inspectors to help them perform calculations in the days before logarithms, computers or pocket calculators. When the House of Commons was electing a new Speaker in 2000, one of the issues on which the candidates were judged was whether or not they would wear a full bottomed wig. One candidate made a revealing comment when he suggested that the Speaker's eighteenth-century costume was desirable because it was 'popular with tourists'.[7] Parliamentary democracy was becoming part of the heritage industry.

Throughout the UK's system of governance, heritage and history determine all sorts of things. It is the Treasury and the Foreign Office that occupy the imposing buildings at the Westminster end of Whitehall, while the Health Department makes do with a hideous concrete edifice a little

[5] Heseltine, M. (2001) *Life in the Jungle*, p.418. Coronet, London.

[6] *Guide to the Public Record Office, Part 1*, p.601/1/2. Public Record Office (1998).

[7] *The Times*, 17 October 2000, p.14. Even the name of the Speaker is now an anachronism; it originally signified that the holder of the office was authorized to 'speak' to the monarch on behalf of the members of the House of Commons.

bit further up the road, slightly more distant from the seat of power. The Office of Science and Technology is crammed into an unimpressive building that is further away still, and which is overshadowed by New Scotland Yard, an imposing (though ugly) modern building which is really a monument to Victorian ideals of expanding the State to create huge bureaucracies like the Metropolitan Police Force.

The structure and geography of government allow no recognition that the modern Treasury, Foreign Office and Police Force depend entirely on science and technology for their operation, or that there is ample evidence that existing systems prevent the proper flow of information, especially scientific information, and that this hinders good government. In the history of mad cow disease, there is no doubt that serious mistakes were made because the structure of government dictated that the Ministry of Agriculture was treated as a second-class ministry.

* * *

Until 2001, the Ministry of Agriculture was at the bottom of the pile in the hierarchy of government. After the General Election in that year, the Prime Minister was so sick of the ministry's performance that he abolished it and replaced it with a Department of the Environment, Food and Rural Affairs. Most of the civil servants from the old ministry were transferred to the new one, and the only change that most of us noticed was that the nameplate on the door of the ministry was replaced with great speed. It remains to be seen whether the new Department will rise higher in the government hierarchy than the old one did.

Traditionally, the Minister of Agriculture has been considered to rank lower than almost any of his Cabinet colleagues.[8] When Nick Brown got the job as Agriculture Minister, he is said to have joked that his patron, finance minister Gordon Brown, must be losing his influence, and commentators saw the appointment as being 'sent to the Siberian pastures of agriculture'.[9] The ministry occupied some of the ugliest and most depressing offices anywhere in London, and many civil servants are said to have regarded it as a graveyard posting.

It is perfectly possible to argue that its low status was well deserved. You may think that farming should make its own way in the world the same as any other industry, and that, if British supermarkets can sell

[8] Although the Minister of Agriculture sat in the Cabinet, and headed a ministry, the role was not deemed to warrant the title of Secretary of State, which is awarded to almost all other Cabinet Ministers with their own departments (except those with individual, historic titles, such as the Chancellor of the Exchequer or the Lord Chancellor).

[9] Rawnsley, A. (2001) *Servants of the People: The Inside Story of New Labour*, p.164. Revised Edition. Penguin, London.

Kenyan runner beans or Danish bacon cheaper than the home-grown equivalents, then that is tough luck on British farmers. Some may even take a harder line on the old ministry and say that it presided over a string of disasters on food safety, on environmental protection, and on agricultural economics, and conclude that we are far better off without it altogether.[10]

But before coming to such a conclusion, we might want to ask *why* the ministry got itself into the position of allowing these disasters to happen. It cannot have been entirely attributable to second-rate ministers or incompetent civil servants. Science was particularly badly treated by the failure of the Ministry of Agriculture, and the public interest in its workings derives from a despair that out-of-date structures and practices can conspire to thwart the good intentions of good people, who cannot all be totally stupid, and at least some of whom must appreciate that science and technology touch every aspect of our lives, every day.

It was not evil intentions or incompetence that caused the crises over mad cow disease and foot-and-mouth disease, the collapse of public confidence over genetically modified foods, misunderstandings over tuberculosis in cattle and wildlife, potential mismanagement of fish stocks in the North Sea, or the scares involving *Listeria* in cheese, *Salmonella* in eggs, and *E. coli* bacteria in meat pies. All of these problems have a single theme in common – their solutions depended on science and on scientists.

The person with prime responsibility for the scientific research that guided policy on these issues was the Chief Scientist at the Ministry of Agriculture,[11] and the person with overall responsibility for coordinating the government's science policy was the Prime Minister's Chief Scientific Adviser. Any sensible person would imagine that these two would be in close contact when a major problem arises but, astonishingly, the archaic structure of UK governance seems to prevent this from being the case.

When a Select Committee of the House of Commons quizzed Dr David Shannon about his role as Chief Scientist at the Ministry of Agriculture, he reported that on many official matters, he felt unable to deal directly with the Chief Scientific Adviser because they were not at the same level of the civil service hierarchy. Dr Shannon said, 'I report to the Permanent Secretary [the senior civil servant of the Ministry of Agriculture] and the

[10] The Prime Minister's decision to abolish the Ministry of Agriculture was certainly fuelled by the widely-held perception that its poor performance during the early stages of an epidemic of foot-and-mouth disease had led to a failure to control the disease as rigorously as would otherwise have been possible.

[11] When the ministry was abolished, its Chief Scientist was simply transferred to the new Department of the Environment, Food and Rural Affairs.

Permanent Secretary talks to Sir Robert May [the Prime Minister's Chief Scientific Adviser], who is at his level [in the civil service hierarchy].'

When asked how the Permanent Secretary could possible deal with the intricacies of detailed issues of science policy when he had 'one thousand and one things on his mind,' Dr Shannon could only reply that he did so because he was 'the boss of the department'. [12]

The Prime Minister's Chief Scientific Adviser gave a slightly different picture of his relationship with the Ministry of Agriculture, claiming that he dealt both with the ministry's Permanent Secretary and with its Chief Scientist. Unlike Dr Shannon's hierarchical view of the civil service, Sir Robert May believed that there was 'not a line of command'.[13] This no doubt reflected Sir Robert's background as an academic researcher, an outsider rather than a career civil servant.

These questions were asked in the context of the handling of the crisis over BSE, in which people died from vCJD, a human form of the illness. With clear hindsight, it seems that at least some of those deaths *could* have been prevented if more effective policies had been put in place early enough. There can be no proof that a different structure, or a different institutional ethos, in the Ministry of Agriculture would have led to such effective policies, but it is obvious that the somewhat antiquated views did not help.

This incident, however, was by no means the most revealing during the debacle over BSE. That honour must go to some of the detailed evidence that was given to the public inquiry set up to determine what had gone so badly wrong that the disease had been allowed to spread throughout the national herd of cattle, and into the human food chain.

* * *

Sir Kenneth Calman was the UK government's Chief Medical Officer during much of the difficult period over mad cow disease, and what he told the official inquiry was almost literally unbelievable.

Sir Kenneth had become worried about the way in which new rules were being enforced in abattoirs and slaughterhouses. In the knowledge that the infective agent was harboured in brain and nerve tissue, the authorities had ordered that the relevant material – known as Specified Bovine Offal – should be completely removed from the carcasses of cattle that were destined for human consumption. The offal was dyed to give it

[12] *Minutes of Evidence of the Science and Technology Committee of the House of Commons*, 2 February 2000, question 127.

[13] *Minutes of Evidence of the Science and Technology Committee of the House of Commons*, 16 February 2000, question 177.

an unmistakable appearance, and there was supposed to be no danger of it reaching the human food chain.

Professor Calman was worried that lapses in the enforcement of the rules 'could have caused Specified Bovine Offal to enter the human food chain' and thought the issue was 'potentially serious'. He wrote a sentence expressing his concerns in a briefing document that was being prepared for government ministers, so that they could take whatever action they thought appropriate in the interest of the electorate. Officials at the Ministry of Agriculture overruled him. They found his statement a 'step too far' and refused to allow the full picture, the whole truth, to be presented to ministers.

The members of the BSE inquiry were so astounded to find out that ministers were not allowed to know the truth that they checked with Sir Kenneth that this document was 'only . . . about advice from officials to Ministers, so Ministers would have a complete picture'. There was no doubt about it. 'Ministers were not being fully informed', said the Chief Medical Officer. Once again, one part of the government apparatus was being prevented from acting in the public interest because of the barriers that have been created by the systems and geography of government.

The Chief Medical Officer is not the only senior figure who lacks genuine authority when it matters. The Chief Scientific Adviser is in the same situation. He told the House of Commons that in at least one government department, the Ministry of Defence, he had 'relatively little impact'.[14]

The depth of the problems associated with this official disorganization became clear when the British government presented its response to the report of the inquiry that had examined how the BSE tragedy could have been prevented.

The report was formally presented to Parliament on 15 February 2001, by Nick Brown, the Minister for Agriculture,[15] which some people may have seen as a mistake in itself. By getting Mr Brown to make the announcement, and his ministry to publish the associated documents, the government *appeared* to be declaring the issue to be one of purely agricultural interest. Although nobody could deny the implications for the UK's agricultural industry, most people were more concerned about human health. With over a hundred people already dead, and scientists saying there could yet be a huge epidemic,[16] people may have had more

[14] *Minutes of Evidence of the Science and Technology Committee of the House of Commons*, 16 February 2000, question 160.

[15] *Hansard* [House of Commons] 15 February 2001, column 485.

[16] *v-CJD in the Future*, Parliamentary Office of Science and Technology, London (2002). [POST Note 171].

faith in the government's words and actions if the issue had been dealt with by the Department of Health.

In fact, it would be completely unfair to blame the government on this issue. It is perfectly obvious to anyone who reads the 100-page document that accompanied the parliamentary statement[17] that no single department or minister could have dealt effectively with the whole issue. It would have been literally impossible for the eighteenth-century system of government to handle this problem that was caused by twentieth-century farming practices, and which required twentieth-century science to solve it.

* * *

The independent team that investigated the BSE tragedy had clearly been astonished at the degree to which the various government ministries had failed to operate in a coordinated way, causing innumerable mistakes during the course of the BSE epidemic and the spread of the new human form of the disease. 'Collaboration between the Ministry of Agriculture and the Department of Health must be maintained', the team said in one of the 16 volumes of the report.

The next volume stressed 'the importance of the agencies working together if threats to human health . . . are to be managed'. 'We reiterate', the team members wrote, 'the importance of continuing to develop closer collaboration'. They mentioned 'joint working at a more detailed level', and said that 'arrangements need to be in place which will facilitate a synchronized approach . . . to common problems of animal . . . or . . . human health'.[18]

The official response from the government spoke only about the Ministry of Agriculture and the Department of Health, totally missing the point that future risks could only be minimized by breaking down the institutional barriers among *all* of the government's ministries and agencies.[19] If the next major problem were an environmental hazard to health (such as a major pollution incident), there would need to be coordination between the Environment and Health ministries; the Ministry of Agriculture would be irrelevant.

'Joined-up government' had been a phrase that was popular for a time among politicians and political commentators, but it seems it did not

[17] *The Interim Response to the Report of the BSE Inquiry by Her Majesty's Government in Consultation with the Devolved Administrations*. The Stationery Office, London (2001).

[18] *Report of the BSE Inquiry*, Volume 1, paragraphs 1269 and 1282; Volume 2, paragraphs 7.79, 7.82 and 7.83. The Stationery Office, London (2000).

[19] *The Interim Response to the Report of the BSE Inquiry by Her Majesty's Government in Consultation with the Devolved Administrations*. The Stationery Office, London (2001).

apply to science policy. The failure to understand that the real issue was more general than just the specifics of BSE was all the more surprising because the Inquiry team had anticipated it and had specifically written, with respect to coordination between the UK government and the separate governments in Scotland, Wales and Northern Ireland, that 'on a more general point, we believe the BSE story has demonstrated the importance of more open sharing of information and research on all topics. This would not only help to meet public concerns but directly facilitate good policy-making'.[20]

In fact, there was a failure by the government and the media to understand the importance of the openness and transparency espoused by most serious scientists. When Dr Ian Gibson, MP, an outspoken champion of scientists, asked the Minister of Agriculture, Nick Brown, whether he knew of the 'attempts by senior civil servants to lean on scientists to suppress their data and interpret them in a manner that fit with the political picture at the time?', the minister refused to comment, saying that there was so much 'evidence of . . . institutional failings and political failings' on other issues that he preferred not to deal at all with Ian Gibson's question.[21] Nick Brown was able to avoid this question only because he could disclaim real responsibility for scientific policies. One of his colleagues, the Secretary of State for Trade and Industry, was nominally responsible for overall science policy.[22] The buck had been neatly passed around within the arcane government structure. It has been successfully buried, and it will never resurface.

* * *

BSE was not the only issue over which the behaviour of people in the Ministry of Agriculture highlighted the effects of a lack of coordination of government policy on science. In the late 1990s, there were at least two other important issues. One was the world's first successful attempt to clone a mammal and the other was the training of future generations of agricultural scientists. In 1996, scientists in Scotland managed to produce a cloned sheep, named Dolly, in a major scientific breakthrough. The news was important not only scientifically, but also for ethical reasons, and the debate about the use of cloning technology will persist for years.

[20] *Report of the BSE Inquiry*, Volume 9, paragraph 17.49. The Stationery Office, London (2000).

[21] *Hansard* [House of Commons] 15 February 2001, column 486.

[22] The Office of Science and Technology is part of the Department of Trade and Industry, and is represented in the cabinet by the Trade and Industry Secretary (the title of President of the Board of Trade was dropped by Michael Heseltine's successors).

But according to Ian Taylor, who was the Minister for Science at the time, the Department of Health and the Ministry of Agriculture were 'umming and ahhing' over whether or not to make an announcement. Their reasons, according to Taylor, were that 'each Department jealously guards its own pitch'. In other words, each department wanted to take any scientific credit that might arise from the announcement, but neither wanted to take the criticism that might come from people who considered cloning to be ethically dubious. In the end, Ian Taylor, whose science portfolio was housed in the Department of Trade and Industry, decided to make the announcement himself, stressing how pleased the government was that a major scientific breakthrough had been made in the UK.[23]

Perhaps the most ridiculous example of how poorly the different government departments coordinate their activities came when the Ministry of Agriculture decided to end its Postgraduate Agricultural and Food Studentship Scheme. The scheme funded bright young researchers to study for their doctoral degrees on agricultural subjects. It was, according to Nick Brown, 'highly regarded' both for the value of its training and for the research that its students produced, and it was seen by many as an essential method of producing the well trained researchers who would deal with the next generation of farming and food problems.

After the Comprehensive Spending Review in 1998, the Ministry of Agriculture found itself financially limited. Its poor image had contributed to a feeling in the Treasury that it did not deserve to maintain the level of public investment that had been put into it in the past. As a result of what Nick Brown later described as a 'difficult Comprehensive Spending Review outcome' for the Ministry, it was decided that the postgraduate scheme should be axed altogether.[24] The reason given by the minister was perfectly legitimate. Postgraduate training was not, of itself, part of the Mministry's specified aims. Its official role was about supporting the rural economy, ensuring the safety of food, and protecting the agricultural environment. Education and training fell within the remit of a different department, the Department for Education.[25]

[23] *The Westminster Hour*, BBC Radio 4, 5 June 2000.

[24] A letter to the Save British Science Society from the Minister of Agriculture, dated 7 June 1999.

[25] The education ministry has changed its name a number of times. In the 1970s, it was the Department of Education and Science. It was later thought that the ministry was 'for' rather than 'of' education, science was later moved to the Department of Trade and Industry, and employment was considered allied to education, so that the ministry became the Department for Education and Employment. When employment matters were moved into a ministry with pensions policy (see Chapter 2), the word 'skills' crept into the name of the Department for Education and Skills. However, in December 2001, the Secretary of State for Trade and Industry twice referred to it as the Department for Education and Science, even though her own department was responsible for science policy. This was yet more proof that among the eternal series of minor changes of departmental names, nobody really knows who is responsible for what.

This reasoning would have been entirely unarguable if there had been sufficient coordination within government to ensure that the Department of Education had taken on the running of the Postgraduate Agricultural Studentship Scheme. But the truth of the matter is not only that the Department of Education did not do so, but that it was not even asked to do so. Officials at the Department of Education were, no doubt, blissfully unaware that an important part of the government's role in education and training was to be discontinued.

Few of us could have any doubts about the reasons why nobody suggested to the Department of Education that it might like to take on the role. It would certainly have said no. Conscious of ever-increasing and unavoidable demands on its own budget, it would have been keen to avoid financial responsibility for something for which someone else had traditionally paid. If the Ministry of Agriculture thinks we need a new generation of farm researchers, it might have said, then the Ministry of Agriculture can pay for them.

The net result, of course, is that the training scheme has vanished, and the role it played will either not be carried out in future, or else it will eventually fall to some other funding agency, which will in consequence have to cut one of its other programmes. None of this would have happened if there were proper and effective coordination of science policy across the government.

* * *

It would be unfair to blame any particular politician or government for the problems of agricultural policy that have been caused by institutional failure based on the historic accident of how a country is governed. It was under the Conservative government of John Major that nobody could sort out who was responsible for Dolly the Sheep, but it was Tony Blair's New Labour administration that casually axed the postgraduate training scheme for agricultural scientists.

The problem with almost any system of government and administration is that its policy is inevitably divided into separate subjects, and individual issues like BSE will always cross the boundaries. That is why none of the existing ministries could have handled the issue any better than the Ministry of Agriculture did. No doubt other ministries would have made different mistakes because their officials would have brought different individual viewpoints to the table.

The only way that most issues can be dealt with in a cross-departmental way is for the Prime Minister either to deal with them personally, or to appoint cross-departmental teams of ministers or officials on an *ad hoc* basis. The first of these solutions is sometimes followed, but it would be unreasonable to expect Prime Ministers to take day-to-day charge of every aspect of government that happens not to fall neatly within the remit of a single department.

Prime Minister Tony Blair did indeed take charge of dealing with an epidemic of foot-and-mouth disease in the spring of 2001, in a very public display of annoyance at the Ministry of Agriculture's failure to get to grips with the outbreak. His pique seems to have been at least partly justified, because the evidence points to a less than brilliant performance by the ministry in the early stages of the epidemic. A previous, devastating, outbreak of foot-and-mouth disease in 1967 had been followed by an inquiry, which had reported that if ever another outbreak occurred, it should be a matter of urgency that expert scientific advice about the likely spread of the epidemic was sought immediately, from whatever quarter the best science could be found. In 2001, a wide range of epidemiological expertise was not drafted in by the Ministry of Agriculture, but had to wait until the Prime Minister took control and sought the assistance of his Chief Scientific Adviser, whose main office is in the Department of Trade and Industry.

This failure was all the more obvious because the message about seeking the best scientific advice had been reiterated much more generally, and very recently, in the Labour government's own guidelines on the use of science in policy making. These had stated less than a year earlier that ministries 'should draw on a sufficiently wide range of the best expert sources, both within and outside government', and had suggested that these 'might include not only eminent individuals, learned societies, advisory committees, or consultants, but also professional bodies, public sector research establishments, lay members of advisory groups, consumer groups and other stakeholder bodies'. To press home the message that ministries should consult anyone who might have something useful to contribute, the guidelines said that ministries should think about 'inviting experts from outside the UK' and that 'experts from other disciplines, not necessarily scientific, should also be invited to contribute, to ensure that the evidence is subjected to a sufficiently questioning review from a wide-ranging set of viewpoints'.[26]

Ministers and civil servants in the Ministry of Agriculture had apparently believed that a sufficiently wide-ranging set of viewpoints was available almost exclusively in-house, even though the Minister had admitted

[26] *Guidelines 2000: Scientific Advice and Policy Making*, p.7. Office of Science and Technology, London (2000).

the previous year that he had been diverting resources away from research on issues other than BSE.[27]

* * *

There is a precedent for dealing with issues that are so all-pervading that they influence every aspect of government policy. The most obvious is that money, which inevitably touches all facets of running a country, has always been handled by a cross-governmental ministry. In the UK, the ministry is called the Treasury, and signals its importance by giving the Prime Minister the extra title of First Lord of the Treasury. Most other countries have similar bodies. The USA, having inherited some of the nomenclature associated with the UK's administration, still appoints a Secretary of the Treasury.

In effect, the UK Treasury predates even the eighteenth-century creation of modern parliamentary government. It has always been a part of government, as a cross-cutting ministry, because money has always been an important part of the policies of every other ministry. It remains the only genuine ministry with a remit that traverses the boundaries of all other departments and, as such, was responsible for the cross-departmental reviews of policy on illegal drugs, local government, care for the elderly, and crime reduction that took place in the spending review of July 2000.

Many parliamentarians believe that the Office of Science and Technology has a similar cross-governmental authority on scientific issues, but they are wrong. On behalf of an all-party group of Members of the House of Commons, Ian Taylor, a former Science Minister under the previous Conservative administration, asked the new minister, Lord Sainsbury, about how he was exercising his 'trans-departmental authority'. The Minister passed the question to his senior official, who said: 'I do not have an authority, I have merely an advisory capacity.'[28]

Even though science and technology are now integral parts of almost every political decision (because they are integral parts of daily life), there remains no effective way of dealing with their daily effects on public policies.

There is no scientific equivalent of the Treasury, with a cross-cutting remit, because successive governments have lacked the imagination to break free from the structure and geography of government that Walpole

[27] *Government Expenditure on Research and Development: The Forward Look*, Volume II, p.81. Fifth Report of the House of Commons Science and Technology Committee, Session 1999–2000. [House of Commons 196-II].

[28] *Minutes of Evidence of the Science and Technology Committee of the House of Commons*, 16 February 2000, question 159.

cobbled together in the 1720s. Occasionally a Prime Minister will rearrange the proverbial deckchairs, as when, in 1992, John Major created what is now the Department of Culture, Media and Sport, or nine years later when Tony Blair abolished the Ministry of Agriculture and invented the Department of Food, Environment and Rural Affairs. But rarely have there been any deeper, fundamental rearrangements that have reflected the enormous social and political changes that have occurred in the last three hundred years.

The UK is by no means unique in the historical inertia that drives the organization of its government. The USA showed the weakness of its own democratic arrangements during the farcical election of George W Bush as President, and its basic structures are only marginally newer than those of the UK, being based on the same principles, but having been given a shake-up in the late eighteenth century during the process of gaining independence.

But the UK is exceptional in the degree to which it adheres to structures that may not be best placed to deal with the scientific realities of the twenty-first century. In the nineteenth century, Anthony Trollope lampooned the system of government by including in his novels a minister with responsibility for the Petty Bag Office. Not much has changed.

The government of the UK still contains someone with the title of Captain of the Gentlemen-at-Arms[29] and, every autumn, Parliament still debates the Clandestine Outlawries Bill of 1676.[30] But when a major food scare, which killed a hundred people and may yet kill many more, was hitting the headlines daily, there was no office with the effective authority to deal with scientific issues across government.

[29] The traditional functions of the Captain of the Gentlemen-at-Arms no longer exist, and the title is now given to the Government's Chief Whip in the upper house of Parliament.

[30] This debate occurs immediately after the government's annual programme is revealed in the Queen's Speech; it serves the useful purpose of asserting Parliament's right to debate whatever it chooses, not just the issues that the government of the day has included in the Speech.

7 Illness, making money and telling us the whole truth

The difficulties of providing high-quality health care and advice about healthy lifestyles have exposed the need for greater openness in dealing with scientific information and scientific uncertainty

THE National Health Service occupies the strangest of places in the hearts of the British people. We all believe we love it, because we think it is a mark of great civilization that everybody should be entitled to free, high quality medical care, irrespective of his or her job, colour, sex or wealth. We love to watch television dramas like *Dr Finlay's Casebook* and *Heartbeat*, which are set in the 1950s and 1960s, and in which friendly local doctors and nurses strive to cure their neighbours of all ills.

But we also expend endless energy and time in denigrating the Health Service because it fails to live up to the high ideals that were set for it when it was founded in 1948. The press is forever finding examples of old women dying on trolleys in hospital corridors, during an eight-hour wait to see a doctor. Professor Michael Joy, a consultant cardiologist with 35 years of experience was so exasperated that he 'finally snapped' when he found forty seriously ill patients on trolleys in the Accident and Emergency department of Saint Peter's Hospital in Chertsey. Speaking in the autumn of 2000, he said that 'in this millennium year, . . . the UK has a standard of emergency care which in my opinion is Third World'.[1]

The truth of course, is somewhere between the two extremes. Like the healthcare systems of all industrialized countries, the National Heath Service is not the idyllic source of care portrayed in television dramas, but nor is it an unmitigated disaster. It truly is an indication of civilization that a nation seeks to make communal provision for the healthcare of those who could otherwise not afford it. When the World Health Organization surveyed the healthcare systems of the world's nations, they found that the UK's system scored extremely highly on a scale of equity.[2] Throughout

[1] *The Times*, 19 October 2000.

[2] *The World Health Report 2000: Health Systems, Improving Performance*. World Health Organization, Geneva (2000).

history, there has never been a society in which the poor, the uneducated and the socially inarticulate have fared as well as those who are rich, the erudite and the well-informed, but the National Health Service works hard to minimize the disparity.

However, the achievements of the Health Service are severely restricted, because demand for such a facility will almost always outstrip supply. As new and better treatments are discovered, more and more finance is needed to provide them. Most of the expensive equipment and drugs that the NHS now uses would have seemed impossible to Aneurin Bevan and his colleagues when their high ideals and boundless energy gave birth to the Service. They cannot have imagined that, within fifty years, it might cost £8,000 every year to provide enough Beta-interferon to alleviate the symptoms of a single patient suffering from multiple sclerosis.[3]

Because we want the best health services, the science of health remains at the top of the public agenda more than almost any other area of research. This interest focuses mostly on the fundamental dilemma of never having enough money to treat everyone's ailments, but always having enough compassion to want to do so. When people are asked to give their own priorities for public investment in science, they always emphasise medical research. In 1996, almost two-thirds of a sample of British adults said that health research should have priority if the government were in a position to reallocate some of the cash that was then being used for military research.

This interest in medical research, and desire for it to be expanded, was made clearer than ever by two photographs in the *The Sun* newspaper.[4] They depicted two Members of Parliament who were leading members of the Science and Technology Committee of the House of Commons. Dr Michael Clark was a Conservative politician who was lucky enough to be given the chairmanship of the committee, even though he represented the Opposition in a Parliament dominated by members of the Labour party, and Dr Ian Gibson was (and is) an outspoken champion of science, who was formerly a biochemist at the University of East Anglia.

These are not the sort of people whose faces we normally expect to see smiling out of the pages of *The Sun*, which is known more for its coverage of the love lives of pop stars and television actors than for its discussions of the work of a parliamentary science committee. Unsurprisingly, this popular and populist newspaper had not covered the committee's

[3] *MS Treatments and NICE*, Parliamentary Office of Science and Technology, London (2001). [POST Note 168].

[4] *The Sun*, 28 July 2000.

previous reports on topics such as *Innovation in the Physical-Based Sciences*, and *The Government's Expenditure on Research and Development*.

The reason for the newspaper's interest in Dr Clark and Dr Gibson was that their committee had chosen to study the science of cancer, a subject of major interest because of reports that cancer patients in the UK were more likely to die than those in other European countries. The government had set a target of reducing deaths from cancer by 20% among those under the age of 75, but the committee felt that more research would be needed to achieve this laudable aim. Michael Clark, Ian Gibson and their colleagues were concerned about 'serious side-effects which commonly limit the treatment [of cancer]' and were keen to exploit 'increasing knowledge and understanding of the molecular basis of regulation of cellular processes and their derangement in the cancer cell'.[5]

The strong attitude that medical research is highly valuable is partly responsible for exacerbating the dilemma of how to maintain a universal health service without bankrupting the nation. New medical research generates new treatments and cures, often for diseases that were previously untreatable, and in some cases were previously undescribed and unnamed. Even when these new cures are inexpensive (which is rarely), they still have to be paid for, and they add to the financial burden of the healthcare system.

Unfortunately, this problem cannot effectively be solved simply by funding health services in a different way. Private health insurance schemes still have to raise enough money to cover the costs of treatment and, when new treatments become available, new money has to be found to pay for them. This is not an ideological opposition to private capital in public services any more than it was ideological to oppose the privatization of the Defence Research Agency, examined in Chapter 5. It is a simple matter of fact.

The public always expects more money to be put into what are known as 'front-line services', which, in the case of health services, means the direct treatment of patients. The Labour government of 1997 was elected on a pledge to reduce the nation's hospital waiting lists, so that 100,000 fewer people would be waiting for treatment by the end of the parliamentary term. This was not a minor policy, buried in the small print of the manifesto, it was one of five specific targets that the party printed onto cards the size of a credit card, and handed out to millions of voters.

[5] *Cancer Research – A Fresh Look*, Volume I, paragraphs 2, 3 and 13. Sixth Report of the House of Commons Science and Technology Committee, Session 1999–2000. [House of Commons 332-I.]

The pledge became increasingly difficult to fulfil, and in October 1999, when the Health Secretary, Frank Dobson, resigned his seat at the government table to fight the election to be Mayor of London, it looked unlikely that the waiting lists would be reduced by 100,000. A new Health Secretary, Alan Milburn, had the opportunity to refocus policies, and was under pressure to drop the commitment.

Mr James Johnson, a surgeon who was a member of the British Medical Association's Council, had accused the government of having 'a completely inappropriate obsession' with numbers on waiting lists. What mattered, he said, was not how many people were waiting, but how many were being treated, and how long they had to wait. Dr David Shubaker felt that priorities were so distorted that 'patients suffering from cataracts and glaucoma are going blind'.[6] But, although Alan Milburn made comments that 'implied criticism of his predecessor', he maintained that 'we will meet the waiting list pledge by the next election'.[7]

In order to achieve this pledge, the government invoked a variety of different policies, including a massive and highly publicized increase in investment for the National Health Service. Chapter 3 explored the way in which the Chancellor of the Exchequer tried to make the funding look even greater than it really was, by announcing three years' worth of financial growth in a single figure.

In their well intentioned attempts to meet a specific pledge, government ministers took some actions which, with hindsight, will come to be seen as perverse. Between their election in 1997 and the time of the next General Election in 2001, ministers cut the research budget of the National Health Service by £58 million.[8] During the 2001 election campaign, the *Birmingham Post* was spearheading a fight to save a local hospital unit from closure and was shocked to learn that research investment had fallen, especially since a league table of research activity had shown, just days earlier, that local scientists were building a 'justified reputation', for, among other things, 'breakthroughs in medical science'.[9]

The government remained resolutely committed to increasing 'front-line' services, and started to pay for a new service called 'NHS Direct', which is a phone line that allows 24-hour access to nurses who can offer advice on whether or not a patient needs to see a doctor. The service was aimed at saving money because some patients could be reassured over

[6] BBC News Online, 7 July 1999.

[7] BBC News Online, 17 October 1999.

[8] *SET Statistics 2000: A handbook of science, engineering and technology indicators*. The Stationery Office, London (2000) [Cm 4902].

[9] *The Birmingham Post*, 26 May 2001.

the telephone, and would not need to access more expensive services. But Dr Hamish Meldrum, a family doctor in Scotland, believed that it would 'ratchet up demand' for services and, hundreds of miles away in London, Dr Laurence Buckman saw the service as a 'Pandora's Box', which was bringing about an 'explosion of patient demand'.[10] As the Health Service tackled waiting lists and expanded the new phone service, its research budget continued to fall.

In theory, there is a simple way of squaring the circle. New drugs and new treatments bring commercial profits for the companies that invent and make them. American pharmaceutical companies sold well in excess of US$130 billion worth of their products in 2000.[11] It is easy to argue that it is these companies that should perform health research, at their own risk, freeing up public money for front-line health services. This line of argument, however, has problems of its own, which can best be illustrated by the row that has emerged over the patents issued to drug companies, particularly in relation to treatments that are essential in beating diseases of the developing world.

* * *

A patent is simply a document awarding an individual or a company an economic monopoly to make or sell a particular invention, and is usually limited for a specific time period. Patents are granted because governments take the view that innovators have the right to recoup the money they invested in developing the new invention. If others could simply copy an invention, they could sell it more cheaply than the originator, because their business would not have to carry the costs involved in the long process of developing the new drug, food, gadget or other invention. This would be a disincentive to inventive companies, which would not bother to carry out research, and to invent new products, if they thought they would lose out financially by doing so. The argument runs that, if a pharmaceutical company spends tens of millions of dollars researching a new drug, it deserves to recoup its outlay before the market is thrown open to competition by other companies, piggy-backing on the investment.

The first patent for an industrial invention was issued in Florence in 1421, and gave Filippo Brucelleschi the exclusive right to manufacture a special barge with hoisting gear used to transport marble.[12] The use of patents to protect inventions has grown enormously since then and, in

[10] BBC News Online, 7 July 1999.

[11] *The 2000 R&D Scoreboard: Company Data*, p.54. Department of Trade and Industry, London (2000).

[12] *Encylopaedia Britannica*, articles entitled 'Patent' and 'Brunelleschi'.

1997, there were 117,000 national patent applications in the UK and 230,000 in the USA.[13] During the twentieth century, patents were taken out on neon light bulbs in 1910, the microwave oven in 1950, and a process for desalinating water in 1984.[14]

Pharmaceutical companies estimate that it now takes around ten years to develop a new drug, and the costs are around £350 million.[15] Not surprisingly, they wish to recoup this investment, and are keen to obtain patents when they develop medicines, and to enforce the rights that those patents give them. Once the patent has expired, or in places where the patent does not apply, other companies can manufacture the drugs cheaply. These copies of the original medicines are known as 'generic' versions of the drug, are chemically identical to the original drug, and are sold much more cheaply because the companies that make them do not incur the costs of research and development.

Some countries, such as India, do not recognize patent protection for pharmaceuticals, so companies can make and sell generic drugs in these countries without waiting for patents to expire. The manufacture of generic drugs is a highly competitive business, and companies strive to keep the costs as low as possible.

Problems arose in southern Africa because the costs of patented anti-AIDS drugs were prohibitively high, with 'triple combination therapy', consisting of three separate drugs, costing around £10,000 per patient per year. Generic versions of the same drugs were available for as little as £250 per person per year.[16] Something like 25 million people in sub-Saharan Africa carry the HIV virus which causes AIDS, and consequently need drug treatment. So by using generic drugs, the health systems of these countries could save £200 billion each year, equivalent to twice the combined national wealth of South Africa, Mozambique, Uganda, Zimbabwe and Botswana.[17]

Unsurprisingly, the governments of these countries felt the need to make anti-AIDS drugs as freely available as possible to their populations, and companies in India, Thailand and Brazil offered to supply South Africa with generic drugs far more cheaply than their brand-name equivalents. The Bombay-based company Cipla Ltd, agreed to sell a cocktail of

[13] *Basic Science and Technology Indicators*, Organisation of Economic Cooperation and Development (2000 edition).

[14] van Dulken, S. (2000) *Inventing the 20th Century: 100 Inventions That Shaped The World*. The British Library, London.

[15] *Access to Medicines in the Developing World*, Parliamentary Office of Science and Technology, London (2001). [POST Note 160].

[16] ibid.

[17] Calculated using data from the website of the World Bank, <www.worldbank.org>.

anti-viral drugs at 'knock-down' prices to the French charity Médicins sans Frontières, on the condition that the charity distributed the drugs free of charge to the poor. 'We are offering drugs at a humanitarian price. This is my contribution to fighting AIDS', said Yusuf Hamied, the chairman of Cipla.[18]

The President of South Africa, Thabo Mbeki, felt particularly strongly that it was unjust for his people to suffer when they were prevented from obtaining drugs that would treat their illness merely because a piece of paper dictated that his government must buy them at expensive prices from the patent holders. So in 1997, South Africa introduced legislation allowing the government to override patent rights in the pharmaceutical sector, in cases where there were good public health grounds for doing so.[19]

The major pharmaceutical companies contested the new law. Led by GlaxoSmithKline, the largest producer of anti-AIDS drugs, 39 companies took the South African government to court, arguing that their patents gave them legal rights. Not surprisingly, the companies were seen by many as chasing profit at the expense of human life. Nelson Mandela, the former President of South Africa and one of the world's most respected people, castigated the big drug companies for 'exploiting' millions of the world's poorest people.[20]

'AIDS is a holocaust against the poor, and responsibility lies with the drug companies who put their profits before their responsibilities', said Zachie Achmat, the head of the South African Treatment Group.[21] The British charity Oxfam threw its weight behind the South African government's claim, and its spokesperson Kevin Watkins said that it was a 'public relations disaster' for the patent-enforcing drug companies.[22]

The big drug companies felt embattled for two reasons. Not only had they been offering cut-price drugs in any case, but it was also less than a year since Thabo Mbeki had publicly questioned the link between the HIV virus and AIDS, and had even gone so far as to say that the anti-AIDS drugs were 'toxic'.[23] He was even accused of letting 'AIDS babies die in pain'[24] and his government was forced to reverse its policy when its popularity fell dramatically.[25] A year later, a woman called Annet Hayman was quoting President Mbeki's former remarks in a court case in which

[18] BBC News Online, 7 February 2001.

[19] The Medicines and Related Substances Control Amendment Act Number 90.

[20] *The Times*, 19 April 2001.

[21] *The Independent*, 7 August 2001.

[22] *The Times*, 19 April 2001.

[23] *The Guardian*, 20 October 2000.

[24] *The Observer*, 20 August 2000.

[25] *The Independent*, 21 October 2000.

she alleged that her husband had been made very ill by the triple combination AIDS drugs.[26] In fact, although the pharmaceutical companies eventually abandoned their case, by February 2002, 'not a single South African [had] been put on a course of . . . drugs as a result of this historic climb down by the drug companies'.[27]

The argument, then, that private companies should be the only large-scale providers of health-related research, cannot be sustained because the economics are complicated and, even in simple cases, can cause major international rows of a financial, legal and diplomatic nature. The public sector has a major role to play, but so does the not-for-profit sector. The UK is lucky that the world's largest medical research charity, the Wellcome Trust, is based in London, and spends more than £1 million a day on British medical research. Britain also has organizations like the Imperial Cancer Research Fund, whose Director, Sir Paul Nurse, won a share of the 2001 Nobel Prize for Medicine for his work on the process of how cells divide.

Together, the various charitable sources mean that the UK has a much higher per capita investment in research from the not-for-profit sector than France, Germany, Japan, or even the USA. But the UK remains surprisingly low in the international league for public sector investment. Per head of the population, the UK government invests about US$150 on research and development each year. The American government spends US$280.[28] It is little wonder the USA has won more than its fair share of Nobel Prizes in the past two decades.

* * *

Public policy on aspects of science related to health carries another major problem – the communication of uncertainty. Doctors and surgeons must find ways of expressing the likelihood that we will recover from a disease after a treatment or operation. Many members of the medical profession have found careful and meaningful ways of telling us, in individual circumstances, that '90% of patients recover fully' from one kind of disease, or that 'about a half of patients live for another five years' after a particular operation. Nevertheless, we must interpret these statistics in our own way because the doctor has no more idea than we do whether we will be one of the 90% who recover, or one of the 10% who do not.

[26] *The Times*, 2 July 2001.

[27] Radford, T. (2002) *Frontiers 01: Science and Technology, 2001–02*, p.44. Atlantic Books, London.

[28] *Basic Science and Technology Indicators* and *Main Science and Technology Statistics*, Organisation of Economic Cooperation and Development (2000 editions).

There is, however, a wider problem for scientists and the public of appreciating more disparate uncertainties. The problem extends far beyond health-related research, but carries particular importance in this area because we expect expert advice to deal sensitively with our need to be reassured that particular foods, activities or treatments are safe, even though we rarely discuss what we really mean by the word 'safe'.

As scientists try to admit their uncertainty, they need to be careful that they are clear about precisely which issues they know about and those of which they are ignorant. This became clear in May 2000, when Sir William Stewart reported to the British government on the potential risks of using mobile telephones. Sir William and a group of eminent scientists had been asked to review the available evidence, following a series of alarming stories in the press. These stories had suggested that the radiation emitted by mobile phones might be harmful to the brains of people using them.

Not surprisingly, William Stewart's review concluded that there was very little evidence either for or against the theory that mobile phones are harmful. Since these telephones had only been invented a few years earlier, it was completely impossible to know much about their long-term effects. Consequently, Stewart and his colleagues used the well-worn phrase that there was 'no evidence' of any adverse health effects from using mobile phones, although there was some preliminary research suggesting that some changes in some biological cells could be caused by mobile phone radiation.

However, the members of the team were careful not to leave themselves open to criticism in the future, since they believed that new research would eventually reveal a deeper understanding of the whole issue. It is entirely possible, they concluded, that children are more at risk than adults because their skulls are thinner and their brains are growing. They had no evidence that any child had yet been harmed by using a mobile phone, but they considered it prudent to be aware of the extra risk that younger people might be running, especially if they were using their phones excessively.

This was a clear attempt to avoid the problems that are usually attached to proclamations that there is 'no evidence' of a problem. These were precisely the words that had led the British government into trouble in the early days of BSE, and William Stewart's group had learned the lesson of that fiasco.

The Times newspaper, however, saw fit to criticize the group for failing to be specific enough in their warnings. Although it thought the precautionary report to be 'broadly judicious', it complained that the document had failed to specify the age at which children's skulls stop thickening and their brains stop growing. It was all very well saying that young people were more at risk than adults, argued the newspaper, but if people were to make practical use of the report, it was essential for them to know the age at which a child grew out of the risk.[29]

Nobody could really argue with *The Times*'s approach. There seems little point in saying that one group of people is safer than others without also giving a hard and fast way for individuals to decide which group they fall into. But therein lies a great problem. William Stewart's review of the risks attached to using mobile phones could easily have said that brains generally stop growing at the age of fourteen, and then parents could have been clear that mobile phones were suitable presents for a fourteenth birthday but not for a thirteenth. However, this approach would have caused its own problems, for the simple reason that every child is an individual and we all develop at different rates.

Everyone knows that some babies start to walk when they are eleven months old, but others do not develop this skill until a number of weeks after their first birthday. We also know that some children grow rapidly when they are in junior school and then stop growing, while their classmates catch up in secondary school. In any aspect of human development, individuals vary in the speed at which they progress and the same is undoubtedly true of the growth of brains and skulls. Some people's skulls will reach their full thickness when they are thirteen and a half, while other children's will not do so until they reach the age of fifteen and a half years. On average, the development of the skull will be complete by the age of fourteen.

But if Sir William Stewart had said that this average age was the point at which the risk of using mobile phones was reduced, then he could be accused of putting at risk any child whose skull was late in developing. His only other option would have been to err on the side of caution, and to say that children should use their mobiles sparingly until they were sixteen, or at whichever age everyone's skull is fully developed. If he had done so, there can be no doubt that he would have been accused of being a part of the 'nanny state', seeking to ban fourteen-year-olds from using mobile phones when most of them would not have been at any greater risk than their parents.

In other words, William Stewart's review group, faced with the inevitable uncertainty of scientific research, was placed in an impossible situation. The members were appointed to review the evidence and make

[29] *The Times*, 12 May 2000.

sensible recommendations but, short of reiterating every detail of the research, their conclusions could only ever be based on averages and generalizations. This means that although their advice could be the best available for the mythical 'average' person, it could *never* be the most perfect advice for real individuals.

The degree to which such advice would suit the general population depends entirely on the degree of variation among the population, in respect of the important feature, which in this case is the speed at which brains and skulls develop. In circumstances where such variation is very limited, advice based on averages might be almost perfect most of the time. By contrast, in situations where such variation is very great, generalized advice is of little use. This kind of variation is the main reason why we all know people who claim that medical advice cannot be right when it warns of the dangers of smoking and fatty diets because they have a granny who has lived to be over one hundred years old on a diet of gin and bacon, complemented with sixty full-strength cigarettes a day.

Of course such people exist but they are rare, and it is foolish for people to create their own rules in life by expecting to perform as well as the best in everything. People who live healthy lives in old age despite their apparently unhealthy eating habits may be lucky for a number of reasons. They might carry a set of genes that gives them the ability to deal with fat and nicotine more effectively than most other people. Alternatively, there might be something else about their lifestyle that has counteracted some of the effects of their unhealthy habits; they might have had an unusually energetic job for fifty years of their life, or have stumbled across the optimal amount of cabbage in their diet to neutralize some of the effects of too much animal fat.

Whatever the reason that an individual can buck the trend, the trend remains. Scientists are still correct to say that it is unhealthy to eat too much butter or to smoke too many cigarettes. Most people will be healthier if they do not smoke than if they do, but a small number of people can get away with smoking. All the scientists can do is to start by giving us a generalization – smoking is bad for you. They can then refine this advice slightly by trying to give some indication of the degree of variation to be found in the human population. The experts might, for example, say that people can probably assume they enjoy some genetic resistance to the effects of smoking if their parents and grandparents had long lives even though they all smoked.

If the experts said this, however, they would need to add a further caveat about genetic variation; not everybody whose relatives resist the consequences of using tobacco will necessarily inherit the genetic material that confers resistance. Genetic inheritance is far from simple, which is why two tall people can sometimes have short children. No scientist can ever be certain about any future eventuality. If scientists are to inform public policy and political debate in a useful way, they must admit not only the fact that they are uncertain about their conclusions, but also the level of their uncertainty. The 'level of uncertainty' about something is just another phrase for probability, and one example of scientists trying to communicate about it is the attempt by television weather forecasters to tell us that there is a 20% chance of rain in London tomorrow, or a probability of 85% that it will be sunny in Washington next week. The degree to which such forecasts are useful is debatable, but they must be preferable to those that give unjustified certainty.

Any attempts to hide uncertainty, or to ignore the variability that causes it, will fail because, by definition, the results will sometimes be utterly wrong. It would be easy for a doctor who knows that 95% of patients recover from a particular treatment to give the impression that everyone recovers. It would reassure an individual patient, and avoid the need to think about the unpleasant possibility of failing to recover; it would also avoid the requirement to explain the doctor's uncertainty. And of course, in 95% of cases, the doctor would get away with this strategy because the patient would indeed survive. But if, during her life, the hypothetical specialist doctor treated two such patients per week for forty years, she would treat a total of perhaps 4,000 people, of whom 5%, or 200 people, would not survive. These 200 people would have been misled.

By the same token, civil servants and politicians should not try to hide their uncertainties about scientific advice, even when the level of uncertainty is small. There will be many issues on which they must pronounce and, if they fail to admit their uncertainty, then on some of those issues the pronouncements will turn out to be wrong. When they find out the truth, the public will feel betrayed.

* * *

A sharp contrast can be made between the way in which politicians and officials tried to inform the public about the health risks to human health of eating beef from cattle affected with BSE, and the way in which the newly formed Food Standards Agency later presented evidence about the health implications of eating lamb from animals that may have been infected with the same disease-causing agent.

Chapter 6 has already explored some of the mistakes that were made over BSE, but it is worth remembering that when the disease was first identified in the 1980s, nobody knew anything about it. Indeed, there remains a small amount of doubt about how it is caused, although a strong scientific consensus now believes the condition is caused by a prion, an unusual form of a normal protein. We are still not completely certain how the prion came to be in the national herd of cattle, with at least one serious scientist believing that it arrived in the UK with antelope that were imported into zoos from Africa.[30]

Although nothing was known about the disease when it first appeared, it was clearly similar to a condition called scrapie which infects sheep, and has been known for well over 200 years. There is no known instance of anyone falling ill from eating scrapie-infected lamb or mutton, so scientists made the imperfect and uncertain guess that the new disease of cattle was likely to behave in similar ways, and that it was probably no more dangerous to eat beef from an infected cow than it was to eat mutton from an infected sheep. As it turned out they were wrong and it is possible for people to catch Creutzfeldt Jakob Disease, the human form of the illness, from eating meat infected with the prion protein. But the scientists had in fact already acknowledged their uncertainty and in 1989, a committee chaired by the eminent zoologist Sir Richard Southwood advised that there was a 'remote' and 'theoretical' risk that people might catch the disease from infected meat.[31]

Ministers and civil servants jumped on the advice that problems were unlikely; they repeatedly assured the public that British beef was 'safe', and that there was 'no risk' from eating it.[32] One Agriculture minister, John Gummer, memorably tried to force his two-year old daughter, Cordelia, to eat a beefburger in front of television cameras, in an attempt to prove that he considered British beef to be safe enough for his own family. The advisability of the photo-call was doubtful, especially since the little girl appeared implacably opposed to the idea of eating the burger. Understandably, the press later described this as a 'disgraceful publicity stunt'.[33]

In March 1996, on the day when the government admitted to Parliament that there might be health risks associated with eating infected beef, Dr R.E. Kendell, the Chief Medical Officer for Scotland, put out a press release saying that 'there remains no scientific evidence that beef itself . . . is unsafe to eat'.[34] Dr Jeremy Metters, the Deputy Chief Medical Officer for England, stressed that 'every effort has thus far been made to

[30] BBC News Online, 18 April 2001.

[31] *Report of the Working Party on Bovine Spongiform Encephalopathy*. Department of Health and Ministry of Agriculture, Fisheries and Food, London (1989).

[32] *Report of the BSE Inquiry*, paragraph 1180. The Stationery Office, London (2000).

[33] *The Daily Mirror*, 20 March 1996.

[34] Press Statement from the Scottish Chief Medical Officer, 20 March 1996.

underline the government's position, based on advice from the Southwood [and other] Committees that the disease is not a risk to humans'.[35]

This was not only inaccurate – the advice was that the disease was *probably* not a risk – it was clearly designed not to inform the public but to give unjustified reassurance, and to protect the meat industry from the disastrous effects of admitting that nobody really knew if their products were safe. After 100 deaths from Creutzfeldt Jakob Disease, and a long and expensive inquiry, it was later realized that 'some statements failed to explain that the views expressed were subject to proper observance of the precautionary measures which had been introduced to protect human health'.[36]

There was a 'public feeling of betrayal'.[37] This had come about because of over-zealous attempts to reassure people. These attempts were hardly surprising given the attitude of most of the media. On 20 March 1996, the day that the government finally admitted there might be a problem, the *Daily Mirror* demanded that 'British beef must be made safe'. Its editor suspected that up to this point, the government had done 'nothing because it did not want to upset the agriculture industry'.[38] There is considerable evidence that some officials behaved in completely unacceptable ways during the crisis surrounding BSE. Chapter 6 has described how a civil servant at the Ministry of Agriculture overruled the Chief Medical Officer, and refused to allow him to keep Cabinet ministers fully informed. But there is also evidence that a failure to appreciate the risks was partly to blame.

The Prime Minister at the time was John Major and in his autobiography, he says that he first realized something was wrong in 1996 when the Deputy Prime Minister, Michael Heseltine, walked into his office with 'that light in his eye and spring in his step which indicated serious trouble ahead'. The government's expert group on cattle disease was to announce 'new evidence of a linkage' between cattle disease and human disease, and John Major says he 'had previously been advised that there was no link'.[39] These are not the words of somebody in the heat of a difficult debate but those of a retired minister, writing and revising his autobiography in discussion with others.

Major is an honest man and he really believes this to be true, even though the Southwood Committee had advised of a potential risk seven

[35] *Report of the BSE Inquiry*, paragraph 1180. The Stationery Office, London (2000).
[36] *Report of the BSE Inquiry*, Executive Summary, Section 5. The Stationery Office, London (2000).
[37] ibid.
[38] *The Daily Mirror*, 20 March 1996.
[39] Major, J. (1999) *John Major: The Autobiography*, p.648. HarperCollins, London.

years earlier. The best scientific advice was that there was 'no evidence' of a link between BSE and human illness, not that there 'was no link'. The only reason the advisers were now able to offer new advice was that ongoing scientific work had been carried out to answer the question of whether the link existed. That would not have been necessary if anyone had been able to tell the Prime Minister that 'there was no link'. Either John Major was actually starting to believe the spin of his own government, or officials were spinning the truth so heavily that he was not being properly informed.

Partly as a result of the tragedy of BSE, a new Food Standards Agency was set up in 2000. Charged with keeping the public properly informed, it has taken the view that just about everything should be put directly into the public domain, including scientific uncertainty. Thus, in August 2001, the Agency announced that 'current plans to promote lamb sales' made it 'timely' to remind consumers of its advice relating to eating lamb. There was, said Sir John Krebs, Chairman of the Food Standards Agency, a 'theoretical risk' that BSE could infect sheep, that it was 'not yet possible to draw conclusions', and that there was need for 'further investigation'. But for the time being, the Agency was 'not advising against the consumption of lamb'.[40]

This advice on the consumption of lamb was rather similar to that given to the government in 1989 regarding beef. But it was handled differently in two ways. It was immediately placed in the public domain, and reported by the mainstream broadcast and print media, and it did not pretend a false level of certainty that lamb was fully safe.

Although James Meikle, a correspondent on the *Guardian* newspaper tried to portray the dangerous aspects of eating lamb by saying 'the spectre of BSE being found in sheep had reared its ugly head again', his subeditor saw that the evidence and advice were much less alarming, and used the headline: 'No immediate need for concern over sheep BSE'.[41] The sale of lamb in the UK did not suffer after the announcement, in sharp contrast to the rapid decline in beef sales in 1996.

By being completely open and admitting both that there could be a health risk and that nobody knew how serious that risk was, the Food Standards Agency was more effective in reassuring the public over the safety of lamb than ministers had been over beef, even though the government's explicit aims had been to reassure consumers and to prevent problems in the agricultural industry. Openness and honesty, especially

[40] *FSA Update on the Risk of BSE in Sheep*, Food Standards Agency Statement, 2 August 2001.

[41] *The Guardian*, 3 August 2001.

about uncertainty, is not just one solution to dealing with scientific advice – it is the only solution that works.

* * *

Since humans evolved conscious thought, we have wanted to predict the future, and people who are believed to have this skill have always been revered. The priestesses of the oracle at Delphi were treated royally, even though their prophecies were renowned for their ambiguity (the prediction that Oedipus would kill his father was 'entirely clear but misleading').[42] Hergé's cartoon character Tin Tin, and Bing Crosby's character in the film *A Connecticut Yankee in the Court of King Arthur* both save their own lives by being able to predict a solar eclipse. Plenty of people try to predict the numbers that will be picked in their local or national lottery, while hundreds of millions of people around the world read their horoscopes in the newspapers every day, and an astonishingly large number believe their predictions will prove accurate.

Some people take prediction very seriously, and gamble huge amounts of other people's money on the outcomes. City traders who play the stock-market are successful only if they can accurately predict which companies will rise in value and which will fall. Employers succeed by predicting which candidates will perform well in a given job, and many students perform well in their examinations by scrutinising past papers and forecasting the questions that are likely to come up again. To know the future is a powerful asset and even when we cannot know precisely what will happen, it is helpful to have confidence that some possibilities are more likely than others.

The same would be true of science if it were possible, although history demonstrates that it is not. Anyone who could have foreseen the importance of electricity or of genetics could have made a fortune, but nobody did predict their importance. Indeed, Gregor Mendel's original work on genetics was ignored by much of the mainstream science community for decades after his death. There are countless examples of clever and well-informed people making statements that appear with hindsight to be utterly stupid, but which at the time seemed perfectly reasonable. It is even reported that Bill Gates once said that 640 kilobytes was enough computer memory for anyone.[43] The computers he makes (and of which he sells millions each year) now routinely come with 15,000 times as much memory.

[42] March, J. (1998) *Cassell Dictionary of Classical Mythology*, p.129. Cassell, London.

[43] Various versions of this story exist, but it was reported by the Chief Executive of the Ordnance Survey at the *Annual Dainton Lecture*, British Library, 28 September 1999.

After the Russians launched the *Sputnik* satellite in 1957, President Eisenhower and his advisers became worried that American science was falling behind the international competition. According to Isidor Rabi, a Nobel Prizewinning adviser, the USA needed to take 'rigorous action' if its technological capability was to avoid being overtaken by the Soviet Union. Eisenhower reorganized the way in which science policy was made, and set up a committee which was charged (among other things) with 'trying to anticipate future trends or developments in the area of science and technology'.[44]

The committee consisted of eminent men such as Hans Bethe, who won the Nobel Prize for Physics, James Killian, who was President of the world-famous Massachusetts Institute of Technology, and William Baker, who was Vice President of the Bell Telephone Company. But while they made many sensible recommendations, and gave American science a boost that left it ahead of the competition for decades to come, they failed to predict the future trends in technology and science. If you think of any piece of science or technology that we use in our homes today, it is a fair bet that Eisenhower's committee did not foresee it. The same is true of the technology used by the military and economic drivers of public policy, even though the group was supposed to place a special emphasis on national security.

The President's committee did not predict the rise of the computer, the use of ultra-precise laser-guided missiles, the huge growth in satellite communications, or the revolution in genetics. Fiction writers probably have a better record of foretelling the future of technology. At least the writers of *Star Trek* had the imagination to predict the invention of the mobile phone. Scientific foresight is not possible on anything but the shortest of timescales. To attempt it is foolhardy. More seriously, it can be against the public interest. It became abundantly clear just how little foresight some politicians have, and how dangerous their attempts can be, when the British House of Commons Science and Technology Committee reported on the ways in which the government used research to inform its policies.

This was the same committee that had produced a report on cancer research to acclaim by the tabloid press. Most of the members of the committee were themselves scientists. The Chairman, Michael Clark, had formerly been an industrial chemist, and other members had held various positions in industrial research laboratories and universities. Lynne Jones had been a researcher at the University of Birmingham and Ashok Kumar

[44] Killian, J.R. (1977) *Sputnik, Scientists and Eisenhower: A Memoir of the First Special Assistant to the President for Science and Technology*, pp.15 and 35. MIT Press, Cambridge, MA.

was formerly a research scientist at British Steel. Not surprisingly, they understood how science works and appreciated that, by its very nature, the process of research is unpredictable and uncertain. The ministers and civil servants whom the Committee interrogated, however, had either failed to understand this at all, or had been forced by political circumstances to act as if it were not true. The outlook of the Minister of Agriculture, Nick Brown, was potentially extremely dangerous.

At the time, the topic that always came up in the context of agricultural research was the tragedy of people suffering long and unpleasant deaths from Creutzfeldt Jakob disease. This is the same condition of the central nervous system that had caused a furore when the government tried to play down fears, ultimately found to be true, that it can almost certainly be contracted as a result of eating beef contaminated with 'mad cow disease'.

Nick Brown was keen to show that he genuinely cared about this important issue, and tried to reassure the members of the parliamentary committee by stressing that in the face of overall budget cuts, he had protected the research programme that was investigating BSE. His ministry had suffered a 'difficult' settlement in the government's spending review, which had meant that it was 'impossible to protect [research spending] entirely from financial pressures on the Department's overall budget'.[45] In fact, his ministry cut its research budget by 40% between 1986 and 2001, by which time it would have cost £88 million per year to restore it to its original level.[46]

This, said Mr Brown, left his ministry 'some way short of our needs', but he proudly asserted that research into transmissible spongiform encephalopathies (the technical name for spongy brain diseases like BSE) had been 'fully protected from reductions' and had been 'increased where necessary'.[47] Nobody could deny the importance of research into BSE, especially since the farming industry was in crisis, and given the horrendous images that had been shown on television of young people who had lost almost all mental and physical co-ordination in the final weeks of their lingering deaths. But the Minister, in his boasts about investing more funds into researching BSE, seemed to have missed what was probably the most important point that the committee of parliamentarians was trying

[45] A letter to the Save British Science Society from the Minister of Agriculture, dated 7 June 1999.

[46] *SET Statistics 2000: A handbook of science, engineering and technology indicators*. The Stationery Office, London (2000) [Cm 4902].

[47] *Government Expenditure on Research and Development: The Forward Look*, Volume II, p.81. Fifth Report of the House of Commons Science and Technology Committee, Session 1999–2000. [House of Commons 196-II.]

to emphasise – being prepared for future potential difficulties is just as important as mopping up after current problems. Prevention is better than cure.

By holding the budget steady for this particular area of research, in the face of overall cuts, Nick Brown must have *reduced* investment in the other areas of his ministry's research programme. To him, it was a matter of current priorities – BSE was a problem *now*, so he must divert resources into dealing with it *now*. Preparing for the future was simply not as important. It is easy to understand why Mr Brown took this approach. The media were constantly clamouring for 'something to be done' about BSE, but they were not clamouring for something to be done about the next problem because neither they, nor anyone else, knew what that next problem was going to be.

In his capacity as a minister, responsible for taxpayers' cash, Nick Brown no doubt took the view that he could hardly put money into researching something that *might*, but probably would not, become a problem.

If he did take this view, then he was wrong. Within a year of his giving evidence to the parliamentary committee, there had been an outbreak of classical swine fever and, more seriously, an epidemic of foot-and-mouth disease that devastated the farming industry, had horrific consequences for animal welfare, and indirectly affected the economy of the whole country by reducing foreign interest in visiting the UK for holidays. The government was forced to set up, and pay for, a scientific inquiry into infectious diseases of livestock, because it had not carried out the appropriate preparatory research when it had the chance. Attempts to predict what will, and what will not be important in the future of science and technology are simply doomed to fail. As Britain's farmers will testify, they are not just against the public interest, they are positively dangerous.

8 Running our railways, the Millennium Dome and the need for failure

Criticism of recent large engineering projects has ignored the essential nature of risk-taking in a successful modern economy

In 1800, a person wanting to travel either a long distance overland, or even a short distance at speed, would have had exactly the same options as someone living five thousand years ago. Animals were still the only source of transport power other than walking, just as they had been for millennia. Different sorts of animals may have been appropriate for different circumstances – a thoroughbred horse for a rich man to ride, a pair of oxen to pull a poor family's cart, or a team of husky dogs to drag an Inuit's sledge – but it was no more possible to move quickly when George Washington wanted to travel around America than it was when Moses needed to cross the desert to the Holy Land.

Since movement has always been essential to human activity and development, it seems surprising that it is only in the past two centuries that any significant progress has been made. We have, however, made up for lost time. In the last two hundred years, change has been so fast that it has sometimes had remarkable consequences.

As Chapter 2 explored, rapid changes in transport technology have had major economic and social consequences. By the twentieth century, many governments around the world had ministries that were more or less devoted to transport policies. Their effectiveness at enhancing our experiences of transportation could reasonably be questioned, given that it is said that the average speed at which people moved around London was the same in 2000 as it was in 1900. Cars can move faster than horses, but only when the infrastructure allows them to. Whether it is strictly true that twenty-first-century cars travel at the same average speed as nineteenth-century horses is barely relevant. That people believe it to be so is enough to make it important.

Not surprisingly, in the last few years of the twentieth century, transport policy rose to heights on the political agenda that it had not seen since its heyday during the railway boom of the 1830s. The privatization of the overground railway system in the UK was much criticized, partly because passengers failed to see significant improvements in service. Throughout the late 1990s, we watched some of the private companies make huge profits, largely because they continued to receive public subsidy, while blaming delays and cancellations on 'the wrong kind of snow' and 'leaves on the line', which seemed to many of us to be poor excuses.

One member of the travelling public complained that he had been forced to stand throughout his journey to work every day for years, because his local train company was so poor at judging how many coaches were required to carry the number of passengers who turned up at his station each morning. Travel was, he wrote to the boss of company, no more comfortable than it had been two thousand years ago. The aggrieved executive replied that, two millennia ago, walking was the only form of travel, and at least his trains provided journeys that were more comfortable than walking to work. The irate traveller sought solace in humour, and pointed out that, in the Bible, it says that thousands of years ago 'the King cometh unto thee . . . riding on his ass',[1] and that this was 'something which I am unable to do on your trains'.[2]

The London Underground system, a great monument to Victorian engineering and the first underground railway anywhere in the world, was a constant source of public argument for years. It became a major political issue when Ken Livingstone was thrown out of the governing Labour Party because of differences on transport policies, but he still managed to win the election as Mayor of London, standing on a platform of opposition to the government's plans for a partial privatization of the system.

In the autumn of 2000, the UK government was again in trouble over its transport policies when a small group of vociferous protestors, angry about the price of fuel for their cars, managed to blockade several oil refineries and cause panic among the population as supermarkets began to fear that, without sufficient fuel to operate their distribution networks, they would run out of supplies.

'Large swathes of the country' found that it was difficult to obtain fuel, and some nurses and ambulances were said to be without sufficient fuel to reach the sick and dying.[3] The people behind the protests, led by a Welsh farmer called Brynle Williams, were said to be a 'rag-tag of farmers and hauliers', but claimed to speak on behalf of the people, although none of them had been elected.[4] Commentators even began to ask whether the Labour government might lose the looming General Election because of

[1] Zechariah IX. 9.
[2] Quoted on *The News Quiz*, BBC Radio 4, 2 November 2001.
[3] *The Independent*, 12 September 2000.
[4] BBC News Online, 3 November 2000.

its seemingly inept handling of the affair when, just weeks earlier, it would have been unthinkable even to suggest that this might be a serious possibility.[5]

* * *

The sorry story of the Railtrack company demonstrates how clearly public policy and political debate about transport can become dangerously disengaged from an understanding of the scientific principles, especially the engineering, on which railways are based. Railtrack was formed when the British government privatized the national railways in the 1990s. From then on (until it was wound up in 2001), the company owned, and was responsible for maintaining the physical infrastructure of, the railway network's rails, platforms and stations.

Public opinion was outraged at what it saw as pathetic excuses for delayed and cancelled trains, and the relatively high fares on some journeys. British national pride was also hurt when, on the same day that the French high speed train system made a record breaking journey from Calais to Marseilles in three and half hours, much slower trains from Edinburgh to London were cancelled because hot weather had warped the rails. Matters were not helped when a Railtrack spokesperson said, 'it's not just about hot days, it's about the combination between hot days and cold nights'. Eventually, Lou Tate, another Railtrack representative, admitted that the company was at fault for failing to lay the tracks properly in the first place.[6]

But more serious criticisms came after what was perceived as a dubious safety record, exemplified by a major train crash at Hatfield in South East England during the autumn of 2000. A shortage of investment was blamed, fairly or unfairly, for the fact that scheduled repairs had not been carried out. Public hatred of the railway managers was especially bitter after the Hatfield crash because it followed just two years after another major accident, outside Paddington station, in which more than thirty people had died, and after which we thought we had been promised the safest railway in the world.

Anyone who knew anything about the railway system was astonished to find out that, before the Hatfield crash, Railtrack did not have a senior engineer on its board of directors. This contrasted greatly with the way the system was run before it was privatized. In the 1960s, British Rail's Chief Mechanical Engineer was 'God-like, with great power, [and] engineering was paramount', according to Tony Roche, a former railwayman

[5] *The Guardian*, 24 September 2000.
[6] *The Independent*, 3 June 2001.

who became the President of the Institution of Mechanical Engineers in 2001.[7]

An inquiry by Lord Cullen, a judge with considerable experience of investigating major disasters, was forced to spell out to railway companies that they were 'expected to demonstrate that they have, and implement, a system to ensure that senior management spend an adequate amount of time' on safety issues such as engineering. Cullen even felt he needed to tell company bosses that the teams responsible should be 'led by the Chief Executive'.[8]

With such a lax attitude to engineering that it needed to be told to have senior engineers involved in its management, Railtrack undoubtedly deserved much of the criticism it received, but for anyone interested in ensuring success for the future, or in making sure that science and engineering are properly put to use for the social and economic benefit of citizens, the question must be asked: How could a public service that depends on engineering be run by a company with no engineering experience?

To be fair to the Conservative government that privatized Railtrack in the first place, its members could hardly be criticized for making the assumption that the company would employ engineers at the highest level. It probably seemed obvious. With Railtrack in trouble, many began to claim that its failure to keep abreast of key issues, such as the engineering of its infrastructure, may have been due to a zeal to make excess profits. According to *The Daily Telegraph*, the company was so 'keen to protect its £1 million-a-day profits . . . [that it] put the squeeze on maintenance costs'.[9]

If greed rather than negligence were indeed the motive for Railtrack's attitude to engineering, the strategy was not successful. Railtrack finally went into receivership in the autumn of 2001 because it had effectively run out of money. The lesson was clear: The public interest is not served – we cannot have an economically viable, safe transport organization – unless public policy puts science and engineering at the heart of the system.

* * *

The railway is by no means the only major engineering system that requires public policies to be based on scientific rigour. In fact, all major civic and public infrastructure requires good engineering. The supply of public utilities such as electricity, water and gas, the construction of

[7] Harris, N. (2001) Putting engineering first. *Rail*, Volume 406, pp.28–33.

[8] *The Ladbroke Grove Rail Inquiry, Part 2*, Summary of Recommendations. The Stationery Office, London (2001).

[9] *The Daily Telegraph*, 22 October 2001.

public buildings, and the operation of bridges and roadways are all based on the principles of engineering, and when something about a public utility changes, there is a need for new and better engineering. Often, those engineering skills go unappreciated, both by the public and by politicians.

The sewerage system of London is amazing because it still works over 130 years after it was first constructed.[10] This is even more surprising than it might seem at first when we realize that when they were built, the Victorian sewers of London had to deal with waste generated by about three or four million people. Today, some seven million people live in Greater London, and the population is boosted to twice this figure during the daytime when millions of extra people travel into the city to work.

* * *

A prime example of the need for engineering in the provision of public utilities comes in the changes that are currently underway in the electricity networks of various countries. The supply of power became a major political theme in the USA in the early part of 2001, dogging the first days of George W. Bush's Presidency. In January of that year, the Governor of California, Gray Davis, declared a state of emergency after a million people in the area around San Francisco Bay had their electricity supply cut off as the State tried to avoid a total collapse of its power systems.

Although Governor Davis believed it was his 'obligation to provide power to the homes and businesses that drive California', he could only guarantee a little over half the electricity demand. The federal government, in the person of US Energy Secretary Bill Richardson, ordered the generators to sell electricity to the State of California, even though they were already owed hundreds of millions of dollars in unpaid bills.[11]

The blackouts even affected major companies in Silicon Valley, the powerhouse of the American high-technology economy, and brought traffic to a standstill. Governor Davis became worried that some companies threatened 'to move manufacturing out of Silicon Valley, even out of State, if we do not see some sort of resolution to reliable energy supplies'.

Officials asked people to eat their evening meals by candlelight. Hollywood was greatly criticized for failing to turn off its Christmas lights, which were still alight even towards the end of January.[12] President Bush was worried by the first deliberate statewide blackouts since the Second World War, and warned that 'the American way of life is under threat'. His Energy Secretary, Spencer Abraham, thought the problems

[10] Hart-Davis, A. (2001) *What the Victorians Did For Us*, p.142. Headline, London.

[11] BBC News Online 18 January 2001.

[12] *The Guardian*, 19 January 2001.

would 'threaten our economic prosperity, [and] compromise our national security'.[13]

Abraham was unquestionably right that we have come to depend on a reliable and consistent power supply in order to live our lives the way we do. But we are also concerned about the environmental effects of burning fossil fuels, such as coal and oil, which are now agreed by almost all the world's scientists as 'likely' to be serious in the coming decades.[14] In addition, we know that supplies of these fossil fuels will not last forever. They are known as fossil fuels because they are the fossilized remains of organisms that died hundreds of millions of years ago, and their supply is limited partly by the difficulties of finding and extracting pockets of oil and gas, but partly by the fact that there is a finite amount under the world's surface. Every time we extract and burn some, we are reducing the reserves.

* * *

Because fossil fuels are limited, expensive and bad for the environment, and because their supply could be erratic, we are seeking increasingly to use alternative sources of power, especially renewable energy sources. Professor David King, the Chief Scientific Adviser to the UK government, believes that in general the nation's science base should fund 'the best science and watch the directions that people choose to follow', rather than trying to predict what will be exciting or useful in the future. 'You have to keep down directed research', he says, in order to use limited resources to fund free-thinking, 'blue skies' research. But in the case of renewable energy research, he thinks that the future problems are so great, and so obvious, that he is prepared to make an exception. He is 'pushing for more research in the area of alternative energy sources . . . There is a need for directed research . . . in this area'.[15]

Renewable energy sources come from natural activities and phenomena such as wind, sunshine, waves and the flow of rivers, as well as human activities such as the incineration of municipal waste and the digestion of sewage sludge. Of the 369 terawatt-hours of electricity generated in the UK in 2000, just 2.8% were generated from these renewable sources, but the government has set a target of 10% by 2010. Parliament passed the Renewables Obligation Order in 2001, forcing the electricity supply

[13] *The Times*, 21 March 2001.

[14] *Summary for Policymakers: A Report of Working Group I of the Intercontinental Panel on Climate Change*, p.2. IPCC, Geneva (2001).

[15] *The Guardian*, 26 October 2001.

companies to sell electricity from renewable sources.[16] We should not underestimate the difficulties for public policy in securing a network that can utilize these new sources of power, and we must appreciate that the required engineering solutions will be extremely challenging.

In most industrialized countries, electricity is generated in a relatively small number of large-scale power stations of various types, and then transmitted around the country at very high voltages via the pylons that cover much of the countryside. It is then distributed at much lower voltages to the majority of households.[17] Although the system is extremely large, it is not inherently particularly complex. Generating more energy from renewable sources and from other small-scale operations will make the situation much more complicated. According to the UK government, the new forms of electricity generation 'cannot be accommodated on the current . . . networks without significant change'.[18]

It will no longer be possible to operate a simple national grid, in which electricity always flows in the same direction from the big power stations to the end users. If a home or small business is generating its own power, perhaps from solar panels, there will be times when it is producing more than it needs, and this excess will need to be fed into the national network to meet demand elsewhere. At other times, the home or business will not be generating enough power for its own needs, and it will need to obtain electricity from the national network. In other words, the distribution network will have to cope with the two-way flow of electricity, something that does not currently occur, and which the existing infrastructure could not handle.[19] This is a very major area of public policy, and its success is entirely dependent on research in the field of electronic and electrical engineering. Most of us may be uninterested in it (it is, after all, rather arcane) but we cannot be disinterested.

* * *

The electricity system, like the Victorian sewers that still deal with our waste, is largely inconspicuous and rarely has any major problems. We tend not to appreciate the engineering beauty of sewers, or even to think about the ensuing challenges for the National Grid. More obvious

[16] *Renewable Energy*, Parliamentary Office of Science and Technology, London (2001). [POST Note 164].

[17] *UK Electricity Networks*, Parliamentary Office of Science and Technology, London (2001). [POST Note 163].

[18] *Technology Status Report: Embedded Generation and Electricity Studies*. Department of Trade and Industry, London (2001).

[19] *UK Electricity Networks*, Parliamentary Office of Science and Technology, London (2001). [POST Note 163].

engineering projects, particularly those that have gone wrong, have been the focus of much more public and political attention.

High-profile projects were deemed necessary for the UK to celebrate the turn of the new millennium on 31 December 1999. The UK has a proud history of celebratory engineering, including the Crystal Palace, built for the Great Exhibition of 1851, or the Dome of Discovery, created a century later at the Festival of Britain. So it was entirely natural that as the new millennium approached, the nation began a whole raft of engineering initiatives, with three huge projects centred in the heart of London – the London Eye, the Millennium Bridge and, of course, the Millennium Dome.

A trip on the London Eye (a 'flight' as the organizers call it) is a truly fascinating experience. To the east, the vast dome of Saint Paul's Cathedral hides among the enormous tower blocks of the City of London. To the west, the Palace of Westminster, Whitehall and Buckingham Palace seem to be crowded together in a small and extremely concentrated centre of historical power. In the distance, the passenger can compare the rising hills of Highgate and Hampstead in the North, with the flat expanse of Greenwich and Woolwich in the south east. At night time, it is possible to get a staggering appreciation of how much electricity must be used to power the millions of tiny specks of light across London.

We can see all this because the London Eye, at 135 metres in height, is the tallest observational wheel in the world. It weighs 1,500 tonnes, but moves effortlessly so that passengers can barely perceive the movement while they enjoy the spectacular views. It is nothing short of fantastic, and it is a triumph of engineering. In fact, it is so novel, and its conception pushed the limits of what was considered possible so far, that it was difficult to raise the wheel to an upright position. The first attempt failed when 'a succession of technical hitches scuppered the operation'. Electronic equipment was rendered useless by interference from satellite dishes in vehicles used by the media to report on the event, and then temporary cables broke free 'with a clang that rivalled that of [nearby] Big Ben'.[20] But within weeks, the problems were sorted, and the wheel was upright.

Another large millennium project where engineering failure was trumpeted by the press was the Millennium Bridge. It has a span of 320 metres, and was an extremely popular project. In June 2000, more than 150,000 people crossed the bridge in the first two days that it was open. The opening ceremony attracted attention because it was attended both by the Queen and by the controversial and newly elected left-wing Mayor of

[20] *The Guardian*, 11 September 1999.

London, Ken Livingstone, whose history suggested he would not relish the thought of sharing a platform with royalty. But 'Red Ken', as the press liked to call him, found that he could not miss the opening of the first new bridge across the Thames for over a hundred years.

Thames bridges have always been centres of great interest, and the people of each era have built new bridges to make their own mark. Until 1176, all London's bridges were wooden and somewhat temporary. In that year, a priest named Peter started to build Old London Bridge, the first great stone arch bridge built in Britain, consisting of 19 pointed arches. It was immortalized in the nursery rhyme 'London Bridge is falling down', and in a children's game of the same name. It was replaced by a new structure of many masonry arches in 1831, which was in turn replaced in the late 1960s. The nineteenth-century structure was so famous that it was moved to Lake Havasu City in Arizona and re-erected as a tourist attraction.

The great civil engineer, John Rennie, built not only the New London Bridge but Southwark Bridge, a construction of three cast-iron arches, and Waterloo Bridge, of three masonry arches. By that time, London also had Westminster Bridge, built in 1750, and Blackfriars Bridge, finished nineteen years later. The Victorians, being great engineers, were keen to make impressive use of their engineering, so it was inevitable that they would also build a new Thames crossing. Tower Bridge, built in 1894, was the last new bridge in London until the Millennium Bridge a century later. Over 75 metres wide, its mechanisms for opening, to allow boats to pass beneath it, were powered by the original steam system until 1976.[21] It was not at all surprising that the people of London should want to celebrate the turn of the new millennium, by carrying on a tradition with such a long and prestigious pedigree.

The Millennium Bridge cost about £18 million to build, with some £7 million coming from the National Lottery. Since its commencement in 1993, British journalists and politicians have loved to criticize the way in which funds from the National Lottery are disbursed. As a consequence of this harshly critical stance towards the committees responsible for spending Lottery cash, the media had a bonanza when the Millennium Bridge had to be closed after just two days. The suspension bridge 'swayed alarmingly under the weight of thousands of people'. Although it was designed to have some movement, those walking across the bridge

[21] *Encyclopaedia Britannica* articles entitled 'Lambeth,' 'London,' 'London Bridge', 'Rennie, John' and 'Tower Bridge'.

complained of 'considerable sway' and 'excessive movement'.[22] The wobbly bridge was declared a disaster.

David Bell, the chairman of the Millennium Bridge Trust said that, although it was 'not true to say the bridge swayed violently', there had been 'two occasions when it moved more than it should have', and he declared the project was a 'big, big disappointment'.[23] A row erupted over who would pay the £5 million bill for repairing the bridge, with the Leader of Southwark Council, Stephanie Elsy, saying that although her Council 'may own the bridge, . . . we are adamant that the residents of the borough . . . will not pay a single penny towards the cost of fixing the problem'.[24] In the end, neither the ratepayers of Southwark nor the general taxpayer contributed anything towards repairing the bridge, although the final deal on who did pay was kept secret.[25]

In all the fuss about the problems with the Millennium Bridge, nobody seemed to point out that, as with the London Eye, doing something new, innovative and different inevitably involves risks. Just as it took two attempts to raise the Eye, so it would take a little longer than expected to make the Millennium Bridge absolutely perfect. Responsible journalists should have looked at the initial failures of past, ultimately successful, projects.

There was nothing new about the knowledge that swaying could result from a large number of people all crossing a bridge at the same time. Some London bridges still retain the plaques that were erected in former years to warn groups of soldiers that they should break step when marching across the bridge in order to prevent the problems caused by hundreds of footsteps all landing on the bridge at precisely the same moment.

Unfortunately, once everyone realized there was a problem with the new bridge, those involved in the project gave unrealistic estimates of how long it would take to fix. David Bell, Chair of the Millennium Bridge Trust, thought the closure would last a few weeks. Tony Fitzpatrick, the chairman of the engineering company, Ove Arup, took a more realistic view. 'When I make a mistake, I fix it', he declared. His company thought the bridge might be closed for six months, meaning that it was unlikely to open again during the millennium year. Tony Fitzpatrick admitted, 'I am an embarrassed man'.[26]

In fact, the Millennium Bridge remained closed for more than a year and a half, much to the delight of the scoffing media. It was not until

[22] *The Guardian*, 12 June 2000.
[23] BBC News Online, 13 June 2000.
[24] *The Times*, 30 November 2000.
[25] *The Evening Standard*, 11 January 2002.
[26] *The Guardian*, 18 November 2000.

January 2002 that 'the last part of the complex, expensive and late-running programme of repairs to the wobbly Millennium Bridge' was put in place, with another month-long series of tests before the bridge was set to re-open.[27]

* * *

The media pundits' criticism of the Millennium Bridge was as nothing compared with their vilification of the Millennium Dome. Unlike almost any other modern building, the Dome instantly became a well-known landmark. Its distinctive shape meant that it was easy for the people in charge of London Transport to create a symbol that unambiguously and uniquely represented the Dome on signposts, without the need for any explanation. The makers of the James Bond films included a scene in *The World is Not Enough* (1999), in which our hero parachuted onto the Dome's exterior.

The reason the Dome came in for a great deal of public criticism is that much of the content was seen to be tacky, inappropriate or plain daft. The entire gamut of newspapers from the *Sun* to the *Financial Times* were hostile, *The Times* ran front-page stories about how disastrous it was on four consecutive days, and even Polly Toynbee, a *Guardian* journalist who had started out as a staunch supporter of the Dome, eventually said, 'Alas, I have to admit, the Dome is a lemon'.[28]

But the structure of the Dome was a piece of engineering of which the UK ought to be proud. It is the largest covered space in the UK, covering 90,000 square metres. Remarkably, that huge structure was made of cables just 32 millimetres in diameter, constructed into a net that basically hung from the distinctive diagonal struts that poked out of the surface. According to Buro Harrold, the engineering company that managed the design, 'in reality, the "Dome" is an enormous stressed cable net structure'.[29]

Perhaps the experience of visiting the Dome would have been more impressive if the organizers had left it empty, and not bothered with the various Zones (the Body Zone, the Money Zone, the Faith Zone and so on) that filled the Dome for its one-year life as a visitor attraction. To stand in that vast empty space and appreciate the skill of the engineering community that created it would probably have been a truly amazing experience.

* * *

[27] *The Evening Standard*, 11 January 2002.

[28] Rawnsley, A. (2001) *Servants of the People: The Inside Story of New Labour*, pp.55, 326 and 328. Revised Edition. Penguin, London.

[29] Information about the construction of the Millennium Dome can be found on the website of Buro Happold <www.burohappold.com>.

The Millennium Bridge, the Millennium Dome and the London Eye are among the science or engineering initiatives that have achieved the highest levels of debate among the public, the media and politicians, and the great majority of that profile and debate has centred on failure.

Little has been made of the clever and successful engineering that has been involved, but that is in the nature of almost all public debate, especially that in which the media takes a lead. Failure, division and humiliation have been the mainstays of politics and journalism for centuries, and not without justification. It is the job of most politicians to keep our few leaders on their toes, and one role of the media is to question and probe public policies. The Watergate scandal of the 1970s was a classic example of how the American press were able, by probing inconsistencies and inaccuracies to expose not just the failings, but also the corruption, of a political leader like Richard Nixon.

The press and political communities have, however, shown a major failure of their own. Too busy criticising the wobbly bridge and the naff Body Zone, political and journalistic hacks have been utterly unsuccessful in appreciating the importance of risk, and in exposing the nation's failure to take risks with science and engineering.

Engineers failed to raise the London Eye on the first attempt because they were doing something new and innovative, and nobody knew whether or not it would work. The Millennium Bridge needed extensive fixing almost immediately because it was new and innovative, and nobody knew whether or not it would work. We spent too much time criticizing the inside of the Millennium Dome, and forgot to recognize that the external structure was new and innovative, and that nobody knew whether or not it would work.

We focused on failure without realizing that failure is a necessary consequence of risk. We cannot both push back the boundaries of human achievement, and live in a cocooned world where nothing can ever fail. This is an important lesson not just for large, public engineering projects but also for managing economic activity and for the funding of the science and engineering research that drives social, environmental and economic progress.

Public policies, partly driven by the agenda of the media, have made us all afraid of taking any risks. Government ministers are asked the question, 'Is this absolutely safe, yes or no?', when anyone who stopped to think about it for a moment would realize this is a silly question. Nothing is absolutely safe, even getting dressed – tens of people die each year in the UK while putting on their socks. But if the minister takes a precautionary approach, and says that an activity (eating genetically modified foods, using a mobile phone or living near an electricity pylon, for example) is not safe, he or she could be giving a highly damaging and misleading impression, because the activity may well be safe in the ordinary sense of the word.

The American humorist, Bill Bryson, gives a good example in his book about walking the Appalachian Trail. The trail runs right the way up the eastern side of the USA from Georgia to Maine and for large parts of its length, hikers can find themselves in remote places where they are unlikely to be heard if they should be attacked and shout for help. Bryson examined the statistics relating to murders on the trail, leaving aside the non-fatal stupidity of people like the woman who smeared her toddler's hand with honey so she could get a photo of a bear licking it off, only to discover that the bear had sharp teeth and had no compunction about eating the child's hand.

There have been nine murders on the Appalachian Trail since it opened in 1937, most of them weird or particularly unpleasant, which makes some people think that hiking the trail is exceptionally dangerous. But considering how many people hike the trail every year (literally millions walk at least some of the way), nine murders in almost seventy years is probably a lower rate than an average small American town. As one of Bryson's friends put it, 'Look, if you draw a two-thousand-mile-long line across the United States at any angle, it's going to pass through nine murder victims'. The simplest understanding of risk dispels the myth that hiking is any more dangerous than, in Bryson's words, sleeping 'in your bed in America'.[30]

* * *

Perhaps the most important role for risk in the twenty-first century is as an essential part of a successful and innovative economy, particularly in a globally competitive environment in which getting smarter is the surest way (indeed the only way) of succeeding. The risks involved are not about real danger; they are about having the nerve to fail.

As Chapter 2 explored, the economy can prosper only if large institutional investors, such as pension funds, choose to invest in companies that take the risk of investing money in research and development. Every so often, one of those companies will discover or invent something that brings substantial benefits in terms of the environment, health or some other factor of life for which we are prepared to pay. But established companies tend to use established research techniques, and a truly vibrant

[30] Bryson, B. (1997) *A Walk in the Woods*, p.197. Doubleday, London.

economy can flourish only if a small proportion of investors' confidence in research is placed with highly innovative, typically small, science-based firms, such as those 'start-up' companies that spring from research ideas in university laboratories.

On the whole, the large investors are extremely cautious with their investments. The money is, after all, not theirs but ours. It is our pension contributions, our savings and our insurance premiums, invested to secure a decent standard of living in later life, or to recompense those of us unlucky enough to have a house burn down, or to suffer some other serious accident. It would be unethical and improper for the pension managers and insurance companies to invest our money in ways that were unlikely to bring decent returns, so it was not particularly surprising when in the autumn of 2001, the high street pharmacist and cosmetic retailer Boots switched all £2.3 billion of its employees' pension savings out of the stock market and into guaranteed bonds, giving fixed rates of interest secured by governments and large companies.[31]

Pension funds in the UK are particularly averse to risk. Between 1996 and 1999, the British Venture Capital Association actually saw a fall of 77% in the amount of money its members were able to raise from UK pension funds, while at the same time there was a modest increase in the amount of venture capital that was available in Britain but which came from foreign pension funds.[32] It seems that, bizarrely, American pension administrators are more keen than their British counterparts to invest in risky ventures in the UK. Institutional investors have always been cautious but, by 1999, even the accountants and actuaries began to complain that the pension funds were becoming too averse to risk. Three major actuarial and accountancy companies supported a call by Tony Blair, the British Prime Minister, that the big pension funds and life insurers should start to invest more in small, high-technology companies that were spinning out of universities as researchers made discoveries with commercial potential.

Even though most such spin-out companies do not succeed in making large sums of money, substantial financial returns can be made by investing in a portfolio of small high-technology firms, because a small number will go on to be so successful that their profits will outweigh the losses associated with the failed companies. 'As a general matter', said the accountants from the companies Bacon and Woodrow, William M. Mercer,

[31] *The Times*, 3 November 2001.

[32] *Report on Investment Activity 1999*. British Venture Capital Association, London (2000).

and Watson Wyatt Partners, 'we support institutional investors making a higher allocation to unquoted securities'.[33]

In fact, 'venture capitalists' were showing much higher returns for their risky investments than the more cautious pension funds were receiving for their investments in large, safe companies, such as those in the *Financial Times* 100-Share Index. In 1997, the average return on cash that had been invested as venture capital was 21%, while a survey covering 1,600 pension funds, and representing three quarters of the British pension industry, found that our pension funds had grown by just 17%.[34]

* * *

Sneering at the Millennium Dome was amusing, and there is no question that much of the criticism was fully deserved. However, the attitude to risk that it demonstrated could potentially be a significant hindrance to a nation's future development, especially in the field of science and engineering.

Most public money is distributed in very cautious ways, for good reason. Taxpayers would not think kindly of a government that took a careless approach to spending their hard-earned cash. But in many areas, particularly in scientific research, managed risk-taking is essential. If we only ever funded research that was sure to work, we would never discover anything that we did not already know or strongly suspected. Many of the world's greatest discoveries have come from experiments that failed. The risks involved are not necessarily dangers, in the sense that people might be harmed, but merely the risk of discovering that one of our theories was wrong or that we must go back to the drawing board because some development has failed.

Alexander Fleming only discovered penicillin when a speck of mould fell into a dish in which he was culturing bacteria. If he had known the mould was there at the start of the experiment, he might well have thought the dish was a failure and thrown it away.[35] The chemist J.C. Swallow described in 1960 how polythene was discovered at ICI's laboratories in Cheshire during the 1930s, merely because the equipment was leaky and dirty.[36]

[33] Press Release issued on behalf of Bacon and Woodrow, William M. Mercer and Watson Wyatt Partners, 6 July 1999.

[34] *1997 Performance Measurement Survey of Independent UK Venture Capital Funds*. British Venture Capital Association, WM Company and Crossroads Management (UK), London (1998).

[35] Roberts, R.M. (1989) *Serendipity: Accidental Discoveries in Science*, p.161. John Wiley & Sons, New York.

[36] Swallow, J.C. (1960) The History of Polythene, in Renfrew, A. (1960) *Polythene – The Technology and Uses of Ethylene Polymers*, 2nd Edition. Iliffe and Sons, London.

Failure of any kind is no longer allowed in modern society, so that it is increasingly difficult to discover success in the wreckage of a failed experiment. Public research money in the UK is now highly controlled by central government in order to avoid failure. 'Prudence' has been a feature of recent public spending reviews, which have sought to ensure that individual ministries sign up to specific outcomes before the Treasury will release their budgetary allocations.[37]

Partly because of an increasingly conservative element in the distribution of existing funds, the government introduced the National Foundation for Science, Technology and the Arts (known as NESTA) in 1998. It is modest in size (it distributes about £10 million a year), but it is ambitious in its aims. Apart from anything else, it is the only government funding body with a specific remit to foster the interfaces between the arts and sciences.

Such an approach may seem bizarre to some but, as Chapter 9 will explore, some scientists see signs of growing public scepticism about their work, and part of the solution is to encourage non-scientists to appreciate science not as something reserved for specialists, but as one more integral part of the national culture in which they can have a stake, and over which they can have some influence. If more people can be genuinely engaged by coupling science with art in imaginative ways, then NESTA will have fulfilled a significant role.

From the purely scientific point of view, there are other reasons for believing NESTA to be an ambitious scheme. An all-party group of Members of the House of Commons see its ambition as being to 'encourage creativity' in individuals, whatever their field of endeavour.[38] Many scientists think that traditional funding sources do not really reward creativity any more. Indeed, some believe that those sources have actually come to dissuade creative science.[39]

Importantly, the members of the Parliamentary Science and Technology Committee have identified a key feature by which it will be possible to tell if NESTA is succeeding in its ambitious plans. 'If NESTA is to succeed overall, it must dare to fail on specific projects'.[40]

[37] *Comprehensive Spending Review Aims and Objectives*. Her Majesty's Treasury, London (1998).

[38] *The National Endowment for Science, Technology and the Arts*, paragraph 5. Second Report of the House of Commons Science and Technology Committee, Session 1998–1999. [House of Commons 472].

[39] *Science Policies for the Next Parliament: Agenda for the Next Five Years*. Save British Science Society, London (2001).

[40] *The National Endowment for Science, Technology and the Arts*, paragraph 6. Second Report of the House of Commons Science and Technology Committee, Session 1998–1999. [House of Commons 472].

That approach will prove difficult for governments to swallow. When Dr Ian Gibson, a member of the Parliamentary Committee, asked about the need for risk in an innovative economy, the Trade and Industry Secretary, Stephen Byers, repeatedly talked about 'calculated risk', and began to give the impression that he thought risk was all very well, as long as there was no chance of failure. Gibson asked Byers if he thought of himself as a risk-taker, to which the Trade Secretary replied, 'Some would say I take too many risks!'[41]

There are those who would agree that Mr Byers took risks with his own career and, within two years, he was in trouble with the press because of various decisions he and his staff had taken. He was accused of 'staggering arrogance' because of an error of judgement in taking a holiday when the railway system for which he was responsible was in chaos.[42] He almost lost his job when he was forced to confess that he gave a misleading television interview.[43] He eventually resigned, after a remarkable period of just a few weeks when he leaked information to the press about matters that were nothing to do with his Department, was lambasted by members of his own party for failing to implement sensible transport policies, and when his recollection of a conversation with the survivor of a serious rail crash differed markedly from the survivor's own account.[44]

But for anyone who wants to see a vibrant economy in which risky science can produce real new innovative benefits, it was ironic that it was while Stephen Byers held responsibility for science policy that the Research Councils increased the degree to which their funds were earmarked for specific schemes,[45] and the complementary Funding Councils proposed that seven new conditions be attached to the money they distributed to universities for research.[46] In publicly funded science, the freedom to fail has been abolished.

[41] *Minutes of Evidence of the Science and Technology Committee of the House of Commons*, 24 February 1999, question 1197.

[42] *The Daily Mail*, 4 January 2002.

[43] BBC News Online, 26 February 2002.

[44] *Financial Times*, 29 May 2002.

[45] 71% of the new money for science announced by the UK Government in 2000 was 'to be directed to . . . genomics, e-science and basic technology', according to *The Science Budget 2001–02 to 2003–04*, Department of Trade and Industry/Office of Science and Technology, London (2000).

[46] *Review of Research: Consultation*. Higher Education Funding Council for England, London [HEFCE 00/37] (2000).

9 What are *they* talking about?

The world of science is improving its engagement with the rest of society, but we still need to get better at talking about the ways in which we organize our scientific endeavours

It has become a truism for scientists to point out that science is now such an integral part of our society that it touches almost every aspect of our lives. This statement has been repeated so often that anyone who works in the field of science must be bored with hearing it. But it is true nevertheless. Previous chapters have explored and illustrated the links between scientific research and the structure of government, transport systems, agriculture, our health, our environment, our jobs and our pensions. But they have also explored how, from time to time, serious problems can arise from a lack of understanding between the world of science and other spheres of human activity.

This problem with communication is unhealthy for us all. It is unhealthy for science, because scientists require the democratic consent of society to legitimize their endeavours. Incomprehension is also unhealthy for all of us as citizens, because we cannot expect to be governed well by people who have little or no understanding of one of the major elements influencing our environment, our society and ourselves. Nor can we expect to control our lives and make the most of opportunities if we do not engage with an important force that drives change in our modern world.

Whether you are a dyed-in-the-wool opponent of scientific change or a rabid techno-geek, it is difficult to argue with the proposition that the world could be improved if science and technology were handled better, especially by governments and public authorities. The examples in earlier chapters have shown that our public policies neither maximize the potential benefits of science, nor minimize the risks associated with the application of scientific discoveries.

A crucial issue for our evolving society is to find ways of bridging the communication barrier, and of integrating science more effectively into our personal lives and public policies, while retaining a healthy scepticism about new technologies, and accommodating a wide range of legitimate viewpoints. All of this must happen while we are all simultaneously allowed to get on with our lives, without having to spend hours every evening doing homework to prepare us for the scientific challenges that tomorrow's news will introduce.

* * *

Some people believe that the public must know and understand more about science, while others think that scientists ought to know and understand more about the rest of the public. Of course, both are right, and both would probably recognize that the word 'public' covers a multitude of definitions. A chemist may think that other chemists are fellow specialists but that biologists are members of the public while, to a government minister, everyone is the public. Sometimes medical doctors are treated as scientists and, at other times, they are not.

Of those who think that the public must know more, some maintain that the way to win the hearts of non-specialists is to fascinate and amaze them with astonishing scientific facts. Others emphasize the importance of understanding the process of scientific discovery, and believe that citizens cannot participate in active democracy on issues like genetically modified crops if they do not understand the experimental basis of the field trials designed to test them.

Among people who believe that it is scientists who must do more to engage with the public, there are those who think that a sunny disposition and a positive, evangelizing style are essential, while others favour a more pragmatic approach, in which researchers admit that they share some of their detractors' concerns. To understand the relative merits of these seemingly conflicting points of view, it is perhaps useful to understand that, although in some cases these new arguments are seeking to solve novel problems, in many ways, they are revisiting age-old issues.

The debate about stem cell research and cloning is an echo of the preface to Mary Shelley's *Frankenstein* in which her husband, the poet Percy Shelley, claimed that 'the event on which this fiction is founded has been supposed by Dr Darwin and some of the physiological writers of Germany, as not of impossible occurrence'.[1] This comment has echoes of the modern views, expressed in the media, that 'designer babies' will be with us in the near future.[2] Moreover, given that Erasmus Darwin had a 'considerable intellect' and was offered the job of personal physician to the king, his views might reasonably have been thought by non-scientists

[1] Shelley, M. (1985) *Frankenstein, or the Modern Prometheus*, p.57. Penguin, London.
[2] *The Daily Express*, 16 October 2001.

to hold some weight, just as concerns are now often voiced by credible scientists.[3]

* * *

Many scientists who lived before the middle of the twentieth century simply took it for granted that they should communicate with a wider audience than just their academic colleagues. It did not seem to them that science was so unusual that it was difficult to talk about, as many people seem to believe now.

The idea, voiced by Lewis Wolpert in the early 1990s, that there is a 'chasm between the two cultures' of scientists and non-scientists[4] was based on C.P. Snow's suggestion that the sciences and the humanities had become alienated from one another.[5] But it would have been seemed bizarre to anyone living in the nineteenth century. The collector of salt taxes in India who, in 1891, wanted to convince local people that they could store large quantities of salt out of doors without it dissolving in the rain, did not assume that Indian peasants could not understand the scientific method. He simply demonstrated 'by practical experiments that heaps of salt thatched, in the open, are almost as safe from wastage as salt stored in [sheds]'.[6] It did not occur to him that the results of empirical experimentation were the preserve of trained 'scientists'.

There is no more effective demonstration that nothing is new in the communication of science than a marvellous painting in the National Gallery in London. It depicts one way in which science was brought into people's homes in the eighteenth century. Called *Experiment on a Bird in an Air Pump*, it was painted in the 1770s by Joseph Wright of Derby. It depicts the eighteenth-century equivalent of television; a man has come into the home to demonstrate science to the occupants, and in the painting he has placed a bird (which looks like a cockatoo) in a glass vessel, and then used a pump to suck the air out, so that the bird is in a partial vacuum. The picture is based on a real event that Wright attended and, on that occasion, the experimenter probably let the air back into the vessel before the cockatoo died, because parrots were valuable in the eighteenth century.

The demonstrator in this picture is one of a long line of people who have, at least to some extent, made their living by spreading the word about science. The list of those scientists who saw it as part of their job to

[3] Porter, R. and Ogilvie, M. (2000) *The Hutchinson Dictionary of Scientific Biography*, Third Edition, Volume I, p.278. Hodder & Stoughton, London

[4] Quoted on the back cover of Jones, S. (1994) *The Language of the Genes*. Flamingo, London.

[5] Gardiner, J. (2000) *The History Today Who's Who in British History*, p.736. Collins and Brown, London.

[6] British Library, India Office, document V/24/3920 (1891–2), p.38.

engage with a wide non-specialist audience is endless. Galileo wrote for a wide audience. In fact, when he published *The Dialogues* (the book that got him into serious trouble with the Catholic church) there were so many advance orders that the book had sold out before the initial print run even reached the bookshops.[7] This publishing success was echoed by Darwin in the 1880s, when he wrote a book about how compost is formed from rotting vegetables.[8] It sold 3,500 copies in the first eight weeks.[9]

The attitude of many historical scientists was, somewhat surprisingly, far more open to engagement with the public than was the attitude of other contemporary professionals. In the nineteenth century, even many politicians did not feel the need to make much effort to interact with a wider public. Lord Liverpool was the longest serving British Prime Minister in history. His Parliamentary career began in 1790 and he came to office as Prime Minister in 1812, relinquishing the premiership shortly before his death in 1828. Strange as it may seem to us, he never gave a single public speech outside Parliament.[10] In a political career lasting 38 years, 15 of them as head of the nation's government, he did not once feel the need to engage with the general public outside Parliament, not even with his own electors in the constituency of Rye.

One of the reasons for Lord Liverpool's reluctance to engage with the masses was that most of them were irrelevant to the fortunes of his political allies. Until the Great Reform Act of 1832, only an elite band of 400,000 landowning men could vote in Parliamentary elections, and they were the only people with whom politicians needed any kind of interface. After several parliamentary reform acts, almost all men over 21 and women over 30 obtained the vote in 1918, by which time the electorate had grown to 21 million. These changes were reflected in the level of political engagement with the public. David Lloyd George, who was Prime Minister in 1918, needed the active support of millions of people, not thousands, and it is impossible to imagine that he could have survived politically without making public speeches around the country. Indeed, he had built his political reputation partly on the basis of provocative speeches he had made, from Newcastle in the North of England to Limehouse in the South, during the bitter campaign in

[7] Sobel, D. (1999) *Galileo's Daughter: A Drama of Science, Faith and Love*, p.231. Fourth Estate, London.

[8] Darwin, C. (1883) *On the Formation of Vegetable Mould by the Action of Earthworms, with Observations on their Habits*. John Murray, London.

[9] White, M. and Gribbin, J. (1995) *Darwin: A Life in Science*. Simon & Schuster, London.

[10] Ramsden, J. (1998) *An Appetite for Power: A History of the Conservative Party since 1830*, p.37. HarperCollins, London.

support of his 'People's Budget' when he was Chancellor of the Exchequer in 1909.[11]

Nineteenth-century scientists were not elected, but they seem to have understood that they relied on the goodwill of the population for a licence to conduct their research. Darwin's *Origin of Species* was not written merely as a scientific monograph, but as a readable book accessible to anyone who wanted to read it, whatever his or her background. It stimulated a great deal of public debate, in which scientists were never afraid to engage in genuinely robust argument with their critics and detractors. Thomas Henry Huxley became known as 'Darwin's Bulldog' because of his robust public backing of the theory of evolution.[12] At a meeting in Oxford, the local Bishop, 'Soapy' Sam Wilberforce, asked Huxley whether it was his grandmother or his grandfather who was descended from apes. Huxley is said to have muttered thanks to God for delivering Wilberforce into his hands, before replying that he was not ashamed of being related to a monkey, but would be ashamed of admitting any connection with a man like Wilberforce who, in Huxley's opinion, was misusing his intellectual skills to obscure the truth.[13]

Even Einstein, whose theoretical work must be among the least accessible to non-specialists, was keen on public communication. His research made the front pages of the *New York Times*, and the public flooded his lectures in such numbers that his students complained that there were no seats left for them. He even wrote a popular book on the theory of relativity. On his first visit to the United States, Einstein wrote that he wanted to correct the 'false opinion widely spread among the general public that the theory of relativity is to be taken as differing radically from the previous developments in physics from the time of Galileo and Newton'.[14]

This attitude appears to have paid off and, by the middle of the twentieth century, it must have seemed to scientists that the general public fully appreciated that everything they did was worthy of unquestioning support. The development of technologies that were beneficial in everyday life, such as antibiotics and electric lighting, fed an increasing belief that science could continue to deliver improvements in our quality of life. In

[11] Ramsden, J. (1998) *An Appetite for Power: A History of the Conservative Party since 1830*, p.208. HarperCollins, London.

[12] Porter, R. and Ogilvie, M. (2000) *The Hutchinson Dictionary of Scientific Biography*, Third Edition, p.518. Hodder & Stoughton, London.

[13] The exact words spoken by Huxley are open to question, and many different versions are now reported in quotation marks. However, the different versions all give the same overall impression, as demonstrated by a comparison of Porter, R. and Ogilvie, M. (2000) *The Hutchinson Dictionary of Scientific Biography*, Third Edition, Volume I, p.518. Hodder & Stoughton, London, and Cadbury, D. (2000) *The Dinosaur Hunters*, p.310. Fourth Estate, London.

[14] Fösling. A. (1997) *Albert Einstein*. Penguin, London.

the 1940s, it was possible for Lancelot Hogben to write a book 'for the large and growing number of intelligent adults who realize that the impact of science on society is now the focus of genuinely constructive social effort'.[15] At the same time, in popular fiction, Nicholas Blake's novels introduced us to Mr Thistlethwaite, a tailor who could say without affectation that 'the scientist is the benefactor of humanity', and to Phyllis Arnold, who thought that the modern technique for surveying public opinion must be a good idea because 'it's scientific after all, isn't it?'.[16]

No doubt the Second World War perpetuated and heightened many people's views that scientific advancement was good for the nation because, as Solly Zuckerman later put it, it was during this conflict that 'Pandora's box opened to reveal to the world the richest store of technological wonders it had ever seen'.[17] It was in the following decades that Harold Wilson promised better times ahead through 'the white heat of technological revolution,'[18] and that President Dwight D. Eisenhower used his valedictory address to speak about the interaction between modern technology and other aspects of the defence and economy of the USA.[19]

* * *

It also appears that it was during the five decades after the Second World War that some parts of the scientific community began to become arrogant and complacent about their relationship with the rest of society. According to one group of scientists, by the 1990s 'many individuals and organizations outside the scientific community believe[d] that the attitudes and behaviour of scientists in the past [had] alienated a large sector of society'.[20] The evidence suggests that they were correct. In 1999, Dr Simon Mitton of Cambridge University Press told a British parliamentary committee that 'we have encountered face to face younger academics who have received instructions not to write books' because it was less important to communicate with a wide audience than to churn out technical

[15] Hogben, L. (1943) *Science for the Citizen: A Self-Educator based on the Social Background of Scientific Discovery*, p.11. George Allen & Unwin, London.

[16] Blake, N. (1940) *Murder with Malice*, Chapters 1 and 5. Harper, London.

[17] Zuckerman, S. (1966) *Scientists and War: The Impact of Science on Military and Civil Affairs*, p.3. The Scientific Book Club, London.

[18] Gardiner, J. (2000) *The History Today Who's Who in British History*, p.848. Collins and Brown, London.

[19] Eisenhower's address is discussed in Zuckerman, S. (1966) *Scientists and War: The Impact of Science on Military and Civil Affairs*. The Scientific Book Club, London

[20] *COPUS looks forward: The Next Five Years*, p.3. Committee on the Public Understanding of Science, London (1992).

articles.[21] More generally, there is among many young scientists a feeling that 'involvement in the popularization of science [is] something that might damage their career[s]'.[22]

A degree of arrogance crept into the scientific community in the second half of the twentieth century. Warren Weaver, who was President of the American Association for the Advancement of Science in 1955, believed that 'the lack of general comprehension of science is . . . dangerous . . . to the public', not because the public had a right to be informed of what was being done with taxpayers' money in a democratic society, but because 'scientists [would] not be given the freedom, the understanding, and the support that are necessary' unless the public made the effort to understand their work.[23]

By the 1980s and 1990s, arrogance was starting to characterize the views of many in the scientific community. In the late 1990s, the astronomer Carl Sagan said of science that members of the public 'don't understand it' but that, if they did, they would support it,[24] while in 1998, an anonymous scientist suggested that the 'public is unable to deal with the [scientific] information in its hands' to Franklin Raines who, as Director of the United States Office of Management and Budget, was the man charged with seeing that the federal taxpayers of the United States obtained value for money.[25]

At the same time, Charles F. Keller said to the physics community that 'the general public . . . will need an appreciation of the sense of wonder . . . that drives humans to learn more about this world', which could almost be interpreted as meaning that non-scientists were not human, and which certainly implied that they had no sense of wonder.[26] Alan Hale, who was in charge of a large part of America's space programme injudiciously used the word 'ignorant' of the public's grasp of science.[27]

Professor Janet Bainbridge, who chaired the UK's Advisory Committee on Novel Foods, had little time for those protesting against genetically modified foods because 'most people do not even know what a gene is'.

[21] *Science and Society: Evidence*, Session 1999–2000, House of Lords Select Committee on Science and Technology. The Stationery Office, London [HL Paper 38-I].

[22] Gregory, J. and Miller, S. (1998) *Science in Public: Communication, Culture and Credibility*, p.1. Plenum Press, New York.

[23] Weaver, W. (1963) Science and people. pp.25–37 in *A Guide to Science Reading*, edited by Hilary J. Deason. Signet, New York.

[24] *The Washington Post*, 30 May 1996.

[25] Greenberg, D.S. (2001) *Science, Money and Politics: Political Triumph and Ethical Erosion*, p.231. University of Chicago Press, Chicago.

[26] Keller, C.F. (1998) untitled letter, *Physics Today*, August 1998.

[27] Quoted in Greenberg, D.S. (2001) *Science, Money and Politics*, p.205. University of Chicago Press, Chicago.

She thought that 'sometimes you have to tell people what's best for them', by which she appeared to mean that people should be told to eat genetically modified foods whether they liked it or not.[28] It is hardly surprising that in the past half century, some non-scientists have started to become irritated by the apparent arrogance of some scientists, or that many scientists have begun to develop the impression that the general public no longer holds them in high esteem, and feel that they need to do something about it.

* * *

In the 1980s, as their budgets were falling, scientists in Britain were so worried about public opinion that they formed a committee which produced a report about the 'public understanding of science', arguing that problems arose because non-specialists simply did not understand enough about science to appreciate its importance. Their report argued that the world would be 'a better . . . place if the public understood more of the scope and the limitations, the findings and the methods of science'.[29]

The same movement has occurred in the United States where Alan Hale, the Director of the Southwest Institute for Space Research believes that 'rampant scientific illiteracy in the general public is . . . one major cause of the current lack of opportunities for scientists'.[30] The man who chaired the British committee, the eminent geneticist Sir Walter Bodmer, produced evidence that 'the more you educate people in science, the more likely they are to support the need for spending more money on science'.[31]

His committee's report spawned a whole new industry, based on the assumption that the public needed to understand more about science. The reasons for that need were often blurred. Although some people stressed the benefits to the average citizen of being able to take part in more debate about scientific issues, many saw the primary cause of the new activities as fostering greater support for science. They predicated their ideas on the assumption that non-scientists were increasingly sceptical about the inevitability that science would produce positive benefits, and thought that the solution was, effectively, some attempt at 'public relations' activities on behalf of science.

[28] Quoted in Monbiot, G. (2000) *Captive State: The Corporate Takeover of Britain*, p.278. Pan Books, London.

[29] *The Public Understanding of Science: Report of a Royal Society ad hoc Group endorsed by the Council of the Royal Society.* The Royal Society, London (1985).

[30] Quoted in Greenberg, D.S. (2001) *Science, Money and Politics*, p.205. University of Chicago Press, Chicago.

[31] Bodmer, W. (1986) *The Public Understanding of Science*, pp.7–8. Seventeeth J.D. Bernal Lecture. Birkbeck College, London.

Scientists were not entirely to blame for the increase in scepticism about their utterances. We are all less deferential to the establishment than we used to be. Those of us who march into our doctors' surgeries with printouts of internet pages about our illnesses, challenging the physician's prescription, are doing something that our grandparents would not have dreamed of doing. But in the modern world, we react angrily against medics like Dr Anthony Hyett who believes that 'those patients who most demand a detailed explanation are those least able to understand it' and that they are 'trying to prove [their] arrogant belief that [they are] so important that [they have] the right to keep other people waiting'.[32]

In one sense, the call for greater 'public understanding of science' has worked. A survey of people in European countries showed that non-scientific members of the general public in the UK actually understand a fair amount about science. When questioned, 79% of the British population understood that in an experiment, such as a trial of a new medicine, the efficacy of the drug could only be properly assessed if patients who took the drug were compared with similar patients who formed part of a 'control group' and who received an inert placebo. By contrast, only 64% of people in Italy and 57% of people in Greece could explain the need for control groups in experimental trials. In the UK, 74% of the population had a realistic grasp of probability, while simple instances of probability were understood by only 45% of the Portuguese and 63% of the Spanish.[33]

However, the views about communicating science that held in the 1980s have not proved entirely robust. The situation is much more complex than the simple idea that when non-scientists understand more about the methods and findings of science, they will inevitably give greater support to scientists, and that such support will, within broad limits, be unconditional. The same survey that showed how much British people understand about science also showed that they are more sceptical about scientific advances than the people of almost any other European country. The Europeans are in general more sceptical than the Americans.

The level of interest in science in the USA is high, with an average level of public interest in new scientific discoveries of 70 on a scale of 0 to 100, and this translates into a less sceptical approach. Among industrialized nations, US citizens report the lowest level of reservation about science

[32] *The Daily Telegraph*, 12 August 1998.

[33] *Science and Society*, p.86. Third Report of the House of Lords Select Committee on Science and Technology, Session 1999–2000. The Stationery Office, London. [HL Paper 38.]

and technology, with the populations of the UK, the Netherlands and Germany showing substantially higher degrees of reservation.[34]

* * *

Much of the scientific community persists in the belief that it is held in unusually low esteem. In November 2001, the UK Treasury published the results of a consultation on the supply of trained researchers in the UK economy. Of 150 organizations involved, including universities, large and small companies, trade associations, individual researchers, pressure groups, government agencies and funding bodies, 'many respondents agreed that the image and public perception of science and engineering was poor'.[35]

Given that scientists are supposed to argue from a basis of evidence, this agreement was remarkably unsupported by hard facts. It is true that some scientific advances are unpopular, or give rise to concerns. Recent examples include genetically modified foods, the handling of food scares, the safety of nuclear power or pesticides, and the need for an experimental cull of badgers to understand the spread of tuberculosis in cattle. It is, however, wrong to interpret these issues as evidence of a universally poor image of science among the wider public.

In fact, there is quite considerable evidence that non-scientists value the work of scientists, and that science has a rather positive image in some quarters. In 1999, Helen Haste, a psychologist from Bath University, compared the image of scientists in the 1970s with that of twenty years later, and found that 'adolescents now think scientists are not only wealthy, but also people who have challenging and rewarding work'. She drew attention to the changing image of scientists in Hollywood films, 'with scientists now presented more realistically in movies from *Jaws*, with Richard Dreyfuss as a marine biologist confronting moral issues, through to *Good Will Hunting*, where Matt Damon gave a sympathetic portrayal of a mathematical genius'.[36]

Whatever British scientists may think, their compatriots respect science. Among the British public, 79% say they are amazed by science and 72% believe it should be funded by the taxpayer even when it has no obvious or immediate benefit.[37] In many surveys about whom we trust,

[34] *Science and Technology Indicators 1998*, p.7.12. National Science Board, Arlington, VA. (1998).

[35] *Review of the Supply of Scientists and Engineers: A Summary of the Responses to the June 2001 Consultation Paper*, p.23. The Stationery Office, London (2001).

[36] *The Independent*, 22 December 1999.

[37] *Science and the Public: A Review of Science Communication and Public Attitudes to Science in Britain.* Office of Science and Technology/The Wellcome Trust, London (2000).

scientists routinely come ahead of the police, all manner of public servants, and even 'the man or woman on the street'.[38]

There remain people implacably opposed to some aspects of science, like the thirteen-year-old girl who wrote a poem about how biologists 'tested' animals 'for fun, and also for the pay'[39] but, in general, science and engineering do not suffer from such a bad image as many scientists seem to believe. Issues like the potential environmental harm caused by genetically modified foods and the dangers of living near mobile phone masts are of genuine concern to everybody, including scientists. People have died from the human form of BSE, or been killed in accidents at nuclear power plants. Science is, like everything else, imperfect, and it would be foolish to pretend that everyone will always be in favour of its advances. But scepticism is not the same thing as hatred. It is not even close.

* * *

Many of the concerns that sceptics raise about modern science are concerned with biological advances. Experimentation with animals is a perennial concern of many, and more recently, there have been worries about the potential environmental consequences of planting genetically modified crops, the ethical issues surrounding the use of human stem cells in medical research, and the possible side-effects of medical procedures such as inoculation with the triple vaccine against measles, mumps and rubella.

Equally powerful are the arguments against the abuse of human genetic material. In October 2001, the *Daily Express* ran a headline about the 'outrage' of a British couple travelling to America to have fertility treatment in order to have a child that was chosen to be a genetic match for its older brother, who needed a bone marrow transplant because he suffered from leukaemia. This 'designer baby' was being 'created to save [its] brother', according to the newspaper, and Nick Harris of the organization known as Life said that the case had 'pushed the boundaries of science too far'. Many scientists agreed and Dr Michael Wilks, the chairman of the British Medical Association's ethics committee, said, 'I don't think helping another child should ever be the primary reason for having a child'.[40]

This example followed a court case in which environmental activists including Lord Melchett, the Director of Greenpeace, had been acquitted of

[38] Worcester, R.M. (1999) *Seeking Consensus on Contentious Issues: Science and Society*, talk to the Foundation for Science and Technology, July 1999, based on a MORI/BMA survey carried out in January 1999, in which roughly 2000 British adults were questioned.

[39] *The Sunday Times*, 14 November 1999.

[40] *The Daily Express*, 16 October 2001.

causing criminal damage after they openly destroyed a field at a farm in Norfolk because it contained a trial crop of genetically modified maize. Their successful defence was that they honestly believed that by destroying the crop, they were protecting other property, including nearby farms. Lord Melchett claimed he had a lawful justification to remove 'genetic pollution'.[41]

A jury of twelve ordinary members of the public agreed with the activists, and although there was no doubt that they had caused the damage, they were acquitted on the grounds that their fears were reasonable, irrespective of whether those fears ultimately turned out to be accurate. In other words, the jurors felt there was some justification for concerns over the *potential* harmful effects of genetically modified foods.[42] In another case, Barbara Charvet and Jim Ridout were acquitted of a charge of spoiling genetically-modified crops, even though they had clearly done it, because the court accepted that their action was a 'principled' stance against a technology that might have effects that are not yet understood.[43] Some evidence suggests that this kind of scepticism about scientific advances is based largely on fear of the unknown, and particularly in a quite proper desire that the activities of scientists should be overseen and regulated by society, generally in the form of society's political representatives. In the UK, 41% of people believe that science now moves too quickly for proper control.[44]

To some degree, the scientific community has become confused because it has started to include under the single heading of 'science communication' three quite separate things. The first activity is simply telling non-scientists about the results of scientific research, the second is offering advice about public policy on contentious issues, and the third is, in return, listening to what the wider public thinks about that advice. Although they are related, these different activities require different skills, generally have different purposes, and are certainly perceived in different ways, both by the scientists themselves, and by those with whom scientists seek to communicate.

* * *

At its simplest level, 'science communication' is about scientists sharing with others the excitement of what they have discovered. In general, they

[41] BBC News Online, 26 September 2000.

[42] *The Independent*, 21 September 2000.

[43] *The Guardian*, 20 November 2001.

[44] *Science and the Public: A Review of Science Communication and Public Attitudes to Science in Britain*. Office of Science and Technology/The Wellcome Trust, London (2000).

share only the 'best bits', a distilled summary of what they have learned rather than the intricate details of how it was learned, and this kind of work tends to exclude all of the failed experiments, and glosses over the gaps in scientists' knowledge.

Proponents of this kind of work can point to some very major successes. When, in 1988, the Cambridge Astronomy Professor Stephen Hawking wrote *A Brief History of Time*, few people believed that a book about the origins of the universe could capture the imaginations of millions of people. But even though some of its concepts are not easy to grasp, the book topped *The Sunday Times* bestseller list for a record-breaking 237 weeks, and has sold many millions of copies around the world.[45]

In broadcasting, the BBC had a similar success with *Walking with Dinosaurs*, a series that used computer graphics and life-like animatronic models to recreate the creatures of 100 million years ago. In the UK, the two soap operas *Eastenders* and *Coronation Street* are by far the most popular television programmes and, most weeks, one or the other tops the viewer ratings. Normally, they can be knocked off the top spot only by an occasional special programme, such as a one-off Christmas show or a televised showing of a major national event. But in the autumn of 1999, *Walking with Dinosaurs* topped the ratings four weeks in a row, relegating the soap operas to second and third places.[46]

Walking with Dinosaurs and *A Brief History of Time* are exceptional, but their popularity is part of a more general feeling that non-scientists enjoy scientific material when it is presented in entertaining and informative ways. The number of popular science books being published is rising faster than the growth in any other kind of books,[47] and major broadcasters like the BBC are producing a growing number of hours of specialist science programmes.[48]

However, it is noticeable that both dinosaurs and astronomy are subjects that, on the whole, do not impinge on the everyday lives of most people. We can be fascinated by these subjects without really caring whether our knowledge is accurate, because we are unlikely to bump into an extinct lizard or to find ourselves in another galaxy in the near future. Our love of popular science should not therefore be confused with a degree of trust or acceptance of science and of scientists on those issues that

[45] Porter, R. and Ogilvie, M. (2000) *The Hutchinson Dictionary of Scientific Biography*, Third Edition, Volume I, p.464. Hodder & Stoughton, London.

[46] Figures from the Broadcasters Audience Research Board, <www.barb.co.uk>.

[47] For example, figures from the Publishers Association show that between 1998 and 1999, there was a 10.8% increase in new science titles, compared to 6.8% for fiction titles, 3.7% for academic and professional books and 7.7% for children's books.

[48] *The 2000 Magnus Pike Lecture*, speech by Caroline van den Brul at the BBC, 6 July 2000.

really do concern us. That is why communicating about science on issues of public policy is a very different activity from merely amazing us with special effects and astonishing facts.

* * *

Governments, and the public they serve, have always needed expert advice in order to decide how to behave, but it seems a reasonable assertion that, in the modern world, scientific advice is even more important than it used to be. A scientific dimension has been an important aspect of a string of issues of national and international importance that have hit the headlines in recent years. From BSE to the efficiency of high-technology weapons in Bosnia and Afghanistan, these issues are much more immediate and important than the lives of dinosaurs or the features of a black hole.

The ancient Greek poet Archilochus said that, in the battle between a hedgehog and a fox, the hedgehog is safe from the predator because although the cunning fox knows many things, the hedgehog knows one big thing.[49] Lord May undoubtedly knows many things (he has been a Professor of Physics in Australia, of Mathematics in America, and of Biology in England), but when (as Sir Robert May) he was the UK government's Chief Scientist from 1995 to 2000, he demonstrated that he knew one big thing. Science can only operate properly in a modern society, and make a positive contribution, if scientists are prepared to tell the truth, and to be open and honest about the scientific issues of the day. Scientific advice must be laid bare before the public, and placed into a robust debate with other issues.

His openness got him into trouble with ministers when he said there was 'not much of a case' for the government's policy of banning beef on the bone to prevent the spread of Creutzfeldt Jakob Disease, and he annoyed the press by saying that the science coverage in the *Daily Express* was 'crap'.[50]

One of the major issues with which Robert May had to deal was a series of protests over the potentially harmful effects of genetically modified crops. What was striking about his behaviour was that he constantly reiterated that although he believed that, in the end, the British public would routinely eat and accept genetically modified foods, he himself had some concerns, particularly about potential harmful effects on the environment.[51] His own scientific background made his remarks all the

[49] Palmer, B.J. (1981) *The Concise Oxford Dictionary of Quotations*, p.7. Oxford University Press, Oxford.

[50] *The Guardian*, 15 August 2000.

[51] *The Guardian*, 1 March 2000.

more powerful – for the previous ten years, he had been a mathematical ecologist studying biodiversity. Robert May's efforts brought about permanent changes in the way that government advisers dealt with this particular issue. In 2001, his successor as Chief Scientific Adviser, Professor David King, said that the public was 'justified to ask what's in it for us to eat these genetically engineered foodstuffs', given that most of us perceived that they were 'untested'.[52]

The stance of May and King was a refreshing change from the attitude taken by some of their colleagues. The environmental campaigner George Monbiot has documented how Professor Ray Baker, when he was Chief Executive of the Biotechnology and Biological Sciences Research Council, was much less open to doubt about the ratio of costs and benefits of genetic modification. Under Professor Baker's leadership, the Council condemned the prestigious medical journal *The Lancet* as 'irresponsible' for publishing a paper by the controversial geneticist Arpad Pusztai, who believed that genetic engineering might endanger human health. According to Monbiot, Ray Baker also dismissed as 'a distraction' a BBC investigation showing that pollen from genetically modified rape plants was being carried more than four kilometres by bees, and on another occasion said that there was 'no risk' of cross-pollination between genetically modified and conventional oilseed rape.[53]

It is this last comment that was perhaps the most damaging. As Chapter 7 showed, few members of the non-scientific public are stupid enough to believe that there is any activity that carries 'no risk', and it was because official spokespeople used this form of words that the public felt betrayed over the issue of BSE, when they discovered that assurances that there was 'no risk' to human health turned out to be untrue. The expensive, thorough and lengthy inquiry into what went wrong over BSE concluded that the poor communication of risks, and of the uncertainty surrounding those risks, was a major factor in exacerbating the crisis.[54]

Scientific advice is all about the communication of risk and uncertainty – Robert May sometimes dealt with his uncertainty in such an open manner that he irritated other scientists, while Ray Baker seems to have left little room for uncertainty in his conclusions, even when others disagreed. Whatever his good intentions, Baker alienated people who did not wholeheartedly agree with him. Chapter 7 examined some of the difficulties of

[52] *EPSRC Newsline*, Winter 2001.

[53] Monbiot, G. (2000) *Captive State: The Corporate Takeover of Britain*, p.292. Pan Books, London.

[54] *Report of the BSE Inquiry*, Executive Summary, Section 5. The Stationery Office, London (2000).

scientists confessing their uncertainty, and the importance of getting it right. The communication of risk is part of the same problem, and presents particular difficulties.

Risk is a compound measure of two different things – the probability that some unfortunate event might happen, and the seriousness of the consequence if it does. Something can be risky for two reasons – either it is very likely to happen or it is likely to cause major problems if it does. An example of the first might be taking the chance that it will not rain if you go for a walk without an umbrella during winter. There is a 'high risk' that you will end up getting wet but, for most people in most circumstances, the consequences are annoying but not life-threatening. The second kind of risk is exemplified by living near an airport. The chances that a plane will fall out of the sky onto your house are very small but, if it happens, there is a 'high risk' that your home will be destroyed and you will probably be killed.

Thus, risk is inherently difficult to communicate partly because it always involves some level of uncertainty, and partly because it is a single measure combining both the probability and the magnitude of an adverse effect. Because it is inherently difficult, and because most of us do not want to hear bad news, politicians in particular have been especially poor at communicating real risks, and the scientific community has, on many occasions, failed to stand up and tell the whole truth.

* * *

The phrase 'public understanding of science' has become almost taboo in many scientific circles. We now talk about 'public engagement'. Many scientists have accepted that members of the non-scientific public were switched off by the apparently arrogant view that any problems could be solved by the non-scientists learning more about the scientists, rather than the other way around.

Almost nobody now believes that telling people about science in general will make them any more appreciative of specific technological advances. Dinosaurs are not ambassadors for genetically modified foods. Nor do we now insist that it is impossible to have a constructive and rational discussion with someone who does not understand the scientific method. A simple demonstration that this belief was unfounded comes from a comparison with the workings of the legal system. Most of us assume that the law is capable of working without the need for us to understand even its most basic functioning.

The result of the 2001 Presidential election in the United States was eventually decided after a lengthy process involving the Supreme Court, the Attorney General of Florida, and the Federal and State courts. If Al Gore had not conceded to George W. Bush, it could have involved the Senate sitting as a court of law. The legal system ground away without anybody much understanding what was going on. Did the Governor of Florida have any authority in the election, and if so how should the Senate and Supreme Court deal with his decisions, given that one of the candidates was his brother? This all followed soon after the impeachment and trial of President Bill Clinton, where someone had to decide what constituted the 'high crimes and misdemeanors' of the American constitution. Clinton even tried to start a debate about the legal definition of the word 'is'.

The American people could take part in a perfectly sensible discussion about whether the legal system was really effective without understanding its impenetrable workings. Likewise, they are perfectly capable of having sensible debates about scientific issues without understanding the intricate workings of the scientific community.

In a similar way, many people in the UK do not know, understand or care about the archaic workings of the House of Commons. They do not understand why a member who wishes to make a point of order during a vote should remain seated but wear a top hat to the attract attention of the presiding officer. This does not mean that these people cannot take part in debate about Parliamentary reform, or that they necessarily lack confidence in the publicly funded system. Nor does it mean that their views are worthless if they express an opinion, based on principle or observation, that the system is failing in some areas, and that some aspects of its workings are undesirable, unfair, unrepresentative, unnecessary, or just silly.

Why then do so many scientists insist that it is essential for us to understand the basic processes of science and that, without this understanding, we cannot contribute effectively to debates about scientific issues? Teaching us a basic understanding of how scientific experiments work is the job of the formal education system, in the same way that schools should inculcate a fundamental grasp of history, geography, languages, the arts, and skills that may prove useful in later life. There is a clear difficulty for a modern education policy in science. Should the curriculum be aimed at training the minority who will form the next generation of specialist scientists, or should it seek to give the majority a basic grounding in scientific ideas and processes, so that we can all take part as active citizens in a democracy? The truth is that it must fulfil both roles.

A survey of industrialized countries showed that, in terms of basic scientific knowledge, the most well informed people were the populations of Denmark, the Netherlands, the United Kingdom, France, Germany and

Italy; in each of these countries, more than 40% of the people were at least moderately well informed. The least knowledgeable people were in Japan and Portugal where less than 20% of people had any claim to being moderately well-informed. Canada and the USA fell somewhere in between with about 25% of the population having a moderate level of scientific knowledge.[55] This relatively low knowledge in the USA seems to coincide with the commonly held view that 'Americans' . . . familiarity with the facts, theories and practices of orthodox science is demonstrably more shaky than that of many other developed countries'.[56]

But there is no reason to suppose that, once the Danes, the British, and the Dutch have gained a good basic grasp of science at school, there should be any expectation of continuing involvement with learning about the scientific method and its application in later life, any more than we expect citizens to be constantly learning more about the legal system, and perpetually keeping abreast of its changes.

It is far more important that we are all alert to the importance of dealing with science in ways that maximize the public interest. It does not really matter whether the vast majority of people know the answers to factual questions about science, or whether they understand the basic processes of science, or even whether they are amazed by the science they see on the television. What matters is that democratic societies know how to structure, to fund, to organize and to apply science in ways that optimize the ratio of benefits and costs. Then the promise of science and engineering can be delivered.

[55] *Science and Society*, p.87. Third Report of the House of Lords Select Committee on Science and Technology, Session 1999–2000. The Stationery Office, London. [HL Paper 38].

[56] Shapin, S. (2001) Guests in the President's House. *London Review of Books*, 18 October 2001, p.3.

10 Concluding remarks

As science, engineering and technology influence our lives in ever more rapidly changing ways, our societies need to find new and effective ways of assessing the advantages and disadvantages of science, and of optimizing the ratio of benefits and costs. In an attempt to do that, various groups of people – such as the media, elected politicians, practising scientists, and pressure groups – are learning to find new and different ways of communicating about science.

Many people have recognised that the ways of the past fifty years are no longer appropriate. They may never have been so (who can say how much better we might have dealt with the problems of global climate change if we communicated more effectively in the past), but now there is no question that the old methods of dealing with science are out-dated. Those of us who are neither in power nor part of the scientific community have already started to recognize that we need to engage more with the interrelationship between science and power. We read more science books than ever, there are more science stories in the news, and more people than ever are involved with environmental organisations, or are taking an interest in whether their pension funds are investing in stocks and shares from high-technology companies.

Yet somehow, society is still failing to deal adequately with the implications of the ways in which science is performed and funded. We still have court cases that rely on an inadequate appreciation of modern statistical and scientific applications, a press that sneers at innovative engineering projects on our behalf, pensions funds that are shy of investing in research-based companies, a deteriorating natural environment, and military research that takes only the shortest-term view of our likely future requirements. Solving these kinds of problems, or deciding that we cannot solve them and must find ways to mitigate and incorporate their effects into new ways of living, will be among the most important tasks for the coming century. Finding solutions and methods of coping will depend on a new level of debate among scientists, the political community, and a wider public. This new debate will not be concerned merely with certainties and amazing facts, but with probabilities, possibilities and predictions.

It is no longer good enough for those people who determine our public policies – politicians, civil servants and political advisers – to pretend that science can be left to a handful of specialists, because it now pervades so many aspects of daily life that it must be factored in to almost every major decision. Increasingly, people are rightly demanding that their elected representatives take seriously their duty to adapt to the changing world. Equally, it is no longer good enough for scientists, engineers and technologists to act as if they form an exclusive priesthood whose language is not and cannot be shared by those who are uninitiated in its secrets. They cannot expect us to make wise choices about science if they make little or no effort to help us. Increasingly, people are demanding that scientists come out of their laboratories and engage with detractors and sceptics. And increasingly, scientists and engineers are finding that this engagement is a positive process.

In fact, science has become just like any other area of public policy, and it is wrong that society does not have more debates about whether we want more or less public funding for science, how we could find better ways of communicating science between different branches of government, whether we want our pension funds to invest a higher proportion of our money in risky, high-technology companies, whether we want drug companies to enforce their patents rigorously even in the developing world, and how to encourage our governments to engage in international collaboration, in order to discover how best to protect the environment. In the coming decades, science and technology will be powerful forces, whether we like it or not. The full benefits of those forces will be realized and the worst problems will be minimized, only if we, as citizens, spend at least as much time and effort involved in the social and political debates surrounding them as we do in debating other areas of public interest.

Index